Edited by
Francesco S. Pavone

Laser Imaging and
Manipulation in Cell Biology

Edited by
Francesco S. Pavone

Laser Imaging and Manipulation
in Cell Biology

WILEY-VCH

WILEY-VCH Verlag GmbH & Co. KGaA

The Editor

Dr. Francesco S. Pavone
European Laboratory for Non Linear Spectroscopy
(LENS)
Polo Scientifico
Sesto Fiorentino, Italy
francesco.pavone@unifi.it

Cover
Two-photon imaging of hippocampal pyramidal
neurons (YFP labelled).

Library of Congress Card No.: applied for

British Library Cataloguing-in-Publication Data
A catalogue record for this book is available from the
British Library.

**Bibliographic information published by
the Deutsche Nationalbibliothek**
The Deutsche Nationalbibliothek lists this
publication in the Deutsche Nationalbibliografie;
detailed bibliographic data are available on the
Internet at http://dnb.d-nb.de.

© 2010 WILEY-VCH Verlag & Co. KGaA,
Boschstr. 12, 69469 Weinheim, Germany

Cover Design   Grafik-Design Schulz, Fußgönheim
Typesetting   Thomson Digital, Noida, India
Printing and Binding   Fabulous Printers Pte Ltd

Printed in Singapore
Printed on acid-free paper

ISBN: 978-3-527-40929-7

Contents

List of Contributors   *XI*

Introduction   *1*
Francesco S. Pavone

Part One   Multiphoton Imaging and Nanoprocessing   *7*

1   **Multiphoton Imaging and Nanoprocessing of Human Stem Cells**   *9*
             Karsten König and Aisada Uchugonova
1.1   Introduction   *9*
1.2   Principle of Two-Photon Microscopy and Multiphoton Tomography   *10*
1.3   Multiphoton Microscopes and Multiphoton Tomographs   *12*
1.4   Endogenous Cellular Fluorophores and SHG Active Biomolecule
             Structures   *14*
1.5   Optical Nanoprocessing   *17*
1.5.1   Principle and Mechanism of Femtosecond Laser Nanoprocessing   *17*
1.5.2   Stem Cells   *18*
1.5.3   Upgrading the Multiphoton Microscope   *20*
1.5.4   Autofluorescence Imaging of Human Stem Cells   *21*
1.5.5   Multiphoton Imaging during Differentiation   *21*
1.5.6   Nanoprocessing   *25*
1.6   Discussion and Conclusion   *28*
             References   *31*

2   ***In Vivo* Nanosurgery**   *35*
             Leonardo Sacconi and Francesco S. Pavone
2.1   Introduction   *35*
2.2   Physical Mechanisms   *36*
2.3   Experimental Setup   *37*
2.4   Subcellular Nanosurgery   *38*
2.5   *In Vivo* Nanosurgery   *41*

Laser Imaging and Manipulation in Cell Biology. Edited by Francesco S. Pavone
Copyright © 2010 WILEY-VCH Verlag GmbH & Co. KGaA, Weinheim
ISBN: 978-3-527-40929-7

2.6   Conclusions   *46*
        References   *47*

Part Two   Light–Molecule Interaction Mechanisms   *49*

**3   Interaction of Pulsed Light with Molecules: Photochemical
          and Photophysical Effects**   *51*
          Gereon Hüttmann
3.1   Introduction   *51*
3.2   Basic Photophysics   *52*
3.2.1   Electronic States of Molecules and the Jablonski Diagram   *53*
3.2.2   Changes between States   *54*
3.3   Bleaching and Excited State Absorption   *57*
3.4   Multiphoton Absorption and Ionization   *60*
3.5   Relevance for Biomedical Applications   *61*
3.5.1   Effectiveness of Pulsed Lasers for Photodynamic Therapy   *61*
3.5.2   Reduction of Photobleaching in Laser Scanning Microscopy   *63*
3.5.3   Super-Resolution by Optical Depletion of the Fluorescent State   *65*
3.6   Conclusions   *67*
        References   *68*

**4   Chromophore-Assisted Light Inactivation: A Twenty-Year
          Retrospective**   *71*
          Daniel G. Jay
4.1   Historical Perspective   *71*
4.2   Family of CALI-Based Technologies   *72*
4.3   Spatial Restriction of Damage   *73*
4.4   Mechanism of CALI   *74*
4.5   Micro-CALI   *75*
4.6   Intracellular Targets of CALI   *75*
4.7   CALI *In Vivo*   *76*
4.8   High-Throughput Approaches   *77*
4.9   Future of CALI   *77*
        References   *78*

5   Photoswitches   *83*
          Andrew A. Beharry and G. Andrew Woolley
5.1   Introduction   *83*
5.2   Synthetic Photoswitches   *84*
5.3   Natural Photoswitches   *89*
        References   *93*

6   Optical Stimulation of Neurons   *99*
          S.M. Rajguru, A.I. Matic, and C.-P. Richter
6.1   Introduction   *99*

6.2   Neural Stimulation with Optical Radiation   *100*
6.2.1   General Considerations   *100*
6.2.2   Effect of Optical Stimulation on Excitability   *101*
6.2.3   Optical Stimulation via Photochemical Mechanism   *101*
6.2.3.1   Activation via Exogenously Added Chromophore   *102*
6.2.3.2   Activation of an Endogenous Chromophore   *102*
6.3   Direct Optical Stimulation of Neural Tissue   *103*
6.3.1   Pulsed Infrared Lasers for Direct Stimulation   *104*
6.3.1.1   Stimulation of Peripheral Nerves   *104*
6.3.1.2   Stimulation of Cranial Nerves   *105*
6.3.1.3   Advantages of Optical Stimulation   *106*
6.3.2   Challenges for Optical Stimulation   *106*
6.3.2.1   Mechanism of Stimulation with Optical Radiation   *106*
6.3.2.2   Safety of Optical Stimulation   *108*
          References   *108*

Part Three Tissue Optical Imaging   *113*

7   **Light–Tissue Interaction at Optical Clearing**   *115*
          Elina A. Genina, Alexey N. Bashkatov, Kirill V. Larin,
          and Valery V. Tuchin
7.1   Introduction   *115*
7.2   Light–Tissue Interaction   *115*
7.3   Tissue Clearing   *120*
7.3.1   Compression and Stretching   *122*
7.3.2   Dehydration and Coagulation   *122*
7.3.3   Optical Immersion   *124*
7.4   Enhancers of Diffusion   *130*
7.4.1   Diffusion through Membranes   *130*
7.4.2   Chemical Agents   *131*
7.4.3   Physical Methods   *132*
7.5   Diffusion Coefficient Estimation   *133*
7.5.1   Spectroscopic Methods   *135*
7.5.2   Optical Coherence Tomography   *138*
7.6   Applications of Tissue Optical Clearing to Different Diagnostic
          and Therapeutic Techniques   *144*
7.6.1   Glucose Sensing   *145*
7.6.1.1   NIR Technique   *145*
7.6.1.2   OCT Technique   *147*
7.6.1.3   Photoacoustic Technique   *147*
7.6.1.4   Raman Spectroscopy   *148*
7.6.2   Tissue Imaging   *149*
7.6.2.1   Confocal Microscopy   *149*
7.6.2.2   Nonlinear Microscopy   *149*
7.6.2.3   Multiphoton Microscopy   *151*

7.6.2.4   Polarized Microscopy   *152*
7.6.2.5   Optical Projection Tomography   *153*
7.6.3   Therapeutic Applications   *153*
7.7   Conclusion   *155*
           References   *156*

Part Four Laser Tissue Operation   *165*

8   **Photodynamic Therapy – the Quest for Improved Dosimetry
              in the Management of Solid Tumors**   *167*
              Ann Johansson and Stefan Andersson-Engels
8.1   Introduction   *167*
8.2   Photodynamic Reactions   *168*
8.2.1   Direct PDT Effects   *170*
8.2.2   Vascular PDT Effects   *170*
8.2.3   Immunological Effects   *170*
8.2.4   Manipulating the PDT Effect   *171*
8.3   Photosensitizers   *173*
8.3.1   Photophysical Properties   *175*
8.3.2   Pharmacokinetics and Tumor Selectivity   *176*
8.4   PDT Dosimetry Models   *177*
8.4.1   Explicit Dosimetry   *179*
8.4.2   Implicit Dosimetry   *181*
8.4.3   Direct Dosimetry   *182*
8.4.4   Biological Response   *184*
8.4.5   Summary of PDT Dose Models   *184*
8.5   Clinical Implementation   *185*
8.6   Where is PDT Heading?   *188*
8.6.1   Novel Applications   *189*
8.6.2   Novel Light Delivery Modes   *190*
8.6.3   Novel Photosensitizer Development   *190*
8.6.4   Novel Implementation of Dosimetry and Dosimetric
           Measurements   *192*
           References   *193*

9   **Laser Welding of Biological Tissue: Mechanisms,
              Applications and Perspectives**   *203*
              Paolo Matteini, Francesca Rossi, Fulvio Ratto, and Roberto Pini
9.1   Introduction   *203*
9.2   Mechanism of Thermal Laser Welding   *206*
9.2.1   Composition of the Extracellular Matrix   *206*
9.2.2   Thermal Modifications of Connective Tissues and Mechanism
           of Welding   *207*
9.2.2.1   Hard Laser Welding   *210*
9.2.2.2   Moderate Laser Welding   *210*

9.2.2.3   Soft Laser Welding   *210*
9.3   Temperature Control in Laser Welding Procedures   *211*
9.3.1   Control Systems of Temperature Dynamics   *211*
9.4   Surgical Applications of Thermal Laser Welding   *214*
9.4.1   Laser Welding in Ophthalmology   *215*
9.4.1.1   Clinical Applications in the Transplant of the Cornea   *215*
9.4.1.2   Preclinical Applications in the Closure of the Lens Capsule   *218*
9.4.2   Laser Welding in Vascular Surgery   *219*
9.5   Future Perspectives   *223*
           References   *226*

           Conclusions   *233*
           Francesco S. Pavone
           References   *242*

           Index   *243*

List of Contributors

Stefan Andersson-Engels
Lund University
Department of Physics
PO Box 118
223 62 Lund
Sweden

Alexey N. Bashkatov
Saratov State University
Research-Educational Institute
of Optics and Biophotonics
410012 Saratov
Russia

Andrew A. Beharry
University of Toronto
Department of Chemistry
80 St. George St.
Toronto, ON M5S 3H6
Canada

Elina A. Genina
Saratov State University
Research-Educational Institute of Optics
and Biophotonics
410012 Saratov
Russia

Gereon Hüttmann
University of Lübeck
Institute of Biomedical Optics
Peter-Monnik-Weg 4
23562 Lübeck
Germany

Daniel G. Jay
Tufts University School of Medicine
Department of Physiology
Boston, MA
USA

Ann Johansson
Munich University Clinic
LIFE Center
Marchioninistr. 23
81377 Munich
Germany

Karsten König
Saarland University
Faculty of Mechatronics and Physics
D-66123 Saarbrücken
Germany

and

JenLab GmbH
D-07745 Jena
Germany

Laser Imaging and Manipulation in Cell Biology. Edited by Francesco S. Pavone
Copyright © 2010 WILEY-VCH Verlag GmbH & Co. KGaA, Weinheim
ISBN: 978-3-527-40929-7

Kirill V. Larin
Saratov State University
Research-Educational Institute of Optics
and Biophotonics
410012 Saratov
Russia

and

University of Houston
Department of Biomedical Engineering
Houston, TX
USA

A.I. Matic
Northwestern University
Feinberg School of Medicine
Department of Otolaryngology
303 East Chicago Avenue
Chicago, IL 60611-3008
USA

Paolo Matteini
Consiglio Nazionale delle Ricerche
"Nello Carrara" Institute of Applied
Physics
Florence
Italy

Francesco S. Pavone
University of Florence
LENS, European Laboratory for Non-
Linear Spectroscopy
Via Nello Carrara 1
I-50019 Sesto Fiorentino, Florence
Italy

Roberto Pini
Consiglio Nazionale delle Ricerche
"Nello Carrara" Institute of Applied
Physics
Florence
Italy

S.M. Rajguru
Northwestern University
Feinberg School of Medicine
Department of Otolaryngology
303 East Chicago Avenue
Chicago, IL 60611-3008
USA

Fulvio Ratto
Consiglio Nazionale delle Ricerche
"Nello Carrara" Institute of Applied
Physics
Florence
Italy

C.-P. Richter
Northwestern University
Feinberg School of Medicine
Department of Otolaryngology
303 East Chicago Avenue
Chicago, IL 60611-3008
USA

Francesca Rossi
Consiglio Nazionale delle Ricerche
"Nello Carrara" Institute of Applied
Physics
Florence
Italy

Leonardo Sacconi
University of Florence
LENS, European Laboratory for
Non-Linear Spectroscopy
Via Nello Carrara 1
I-50019 Sesto Fiorentino, Florence
Italy

Valery V. Tuchin
Saratov State University
Research-Educational Institute
of Optics and Biophotonics
410012 Saratov
Russia

Russian Academy of Sciences
Institute of Precise Mechanics and
Control
410028 Saratov
Russia

Aisada Uchugonova
Saarland University
Faculty of Mechatronics and Physics
D-66123 Saarbrücken
Germany

G. Andrew Woolley
University of Toronto
Department of Chemistry
80 St. George St.
Toronto, ON M5S 3H6
Canada

Introduction

Francesco S. Pavone

Since the development of nonlinear laser imaging tools, such as the two-photon technique [1] for example, many technological advancements have been made in the field of microscopy and, more generally, imaging. It took more than 60 years to move from the discovery of the two-photon interaction [2] to its exploitation in microscopy. Since the 1990s, an exponential growth of publications in the field of microscopy (Figure 1) has led to the introduction of the two-photon technique in the laboratories of many researchers worldwide.

Since the first interaction schemes, where all photons were accumulated and collected on the detector after the laser irradiation (integration mode), other kinds of investigation modes have been developed, based, for example, on the lifetime response of the fluorescent molecule (fluorescent lifetime microscopy), on the spectral behavior of fluorescence emission (multispectral two-photon emission), or on the ability of the illuminated molecule to double the frequency of the coherent excitation due to its nonlinear susceptibility (second- and third-harmonic generation microscopy).

Further developments in microscopy have led to other nonlinear interaction schemes such as coherent anti-Stoke Raman spectroscopy (CARS) [3] (Figure 2) and resonant Raman scattering [4].

The nonlinear characteristic of the interaction of pulsed light with a molecule has also led to applications that are useful in increasing the resolution below the diffraction limited barrier [5].

All these imaging tools, together with well-developed photon based technology, such as confocal microscopy, have enlarged the field of applications in biological imaging of molecules, cells, and tissues.

Consequently, the new frontier of cell biology imaging has moved from a fixed cell to a living cell with the advent of the laser and more sensitive wide-field fluorescent microscopes. The advent of confocal microscope has improved the axial resolution, while the application of multiphoton processes has finally permitted the study of cell biology in tissues and, consequently, in living organisms, as well as allowing optical manipulation [6].

Laser Imaging and Manipulation in Cell Biology. Edited by Francesco S. Pavone
Copyright © 2010 WILEY-VCH Verlag GmbH & Co. KGaA, Weinheim
ISBN: 978-3-527-40929-7

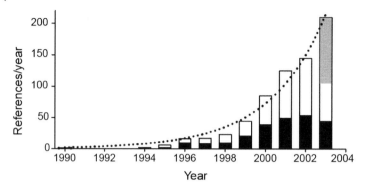

Figure 1 Publications on two-photon microscopes in recent years.

In all cases, a common approach in these methodologies was based on the same concept: labeling the cell with fluorescent probes and observing the topographical changes under some circumstances. Some important characteristics have been measured such as the compartmentalization of some molecules in the cell, their diffusive dynamics, the interaction with other partner molecules, their response to external stimuli, either pharmacological or electrical for example, and so on.

A big step forward has been the study of the nonlinear interaction of light with molecules. The unwanted effect of photobleaching (Figure 3), which generally limits the acquisition time, has turned out to be useful in many applications.

First, a relationship between the photobleaching and the photodamaging energy pathway has shown the possibility of perturbing the chemical state of species by means of a photochemical effect or by creating a plasmon field that perturbs the molecule itself.

A new method of performing cell biology was born: with respect to the "passive" method of "illuminating and observe," the laser has been used for the first time to perturb and observe the reaction of the system.

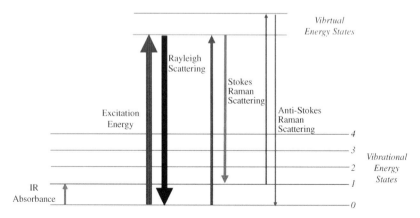

Figure 2 Raman and CARS interaction schemes on molecular energy levels.

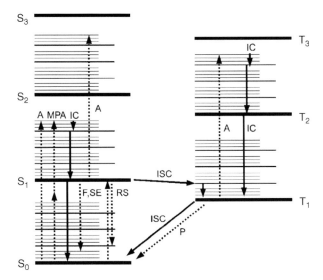

Figure 3 Energy level pathway involved in photobleaching.

This has been applied to single molecule both *in vitro* and in living cells to denaturate molecules, break or move subcellular structures, or create pores into the cell membrane.

The advent of this new cell biology method, a sort of interacting investigation scheme, has led me to gather a team of experts on the topic of "laser imaging and manipulation in cell biology."

My intention is to bring to the attention of the reader a broad panorama of applications from one-photon to multiphoton interaction schemes, from the single living cell *in vitro* to the cell in its physiological environment, like a tissue, where the laser was used to manipulate the sample.

The common aspect bridging all these arguments is based on the "photobiology" of the molecular interaction with light. Even in a tissue, all origins of macroscopic phenomena lie on a molecular base mechanism of interaction with light.

Starting from this consideration, it was easy to present a similar aspect based on a light–molecule interaction mechanism in imaging or manipulation of both cell and tissue.

Following this scheme, the first part of the book is dedicated to the basics of multiphoton imaging and nanoprocessing in cells. The second part focuses on the light–molecule interaction mechanisms. The third part broadens the field of imaging and manipulation from cell to tissue, illustrating tissue imaging applications. Finally, the fourth part describes light manipulation applications on tissues.

The invited authors have made significant contributions to these fields and to applications.

In particular, Professor Karsten Köning has been a pioneer in work on the multiphoton processing of cells [7]. He was in fact one of the first researchers to demonstrate the possibility of using multiphoton interactions to process a cell by

(a)

(b)

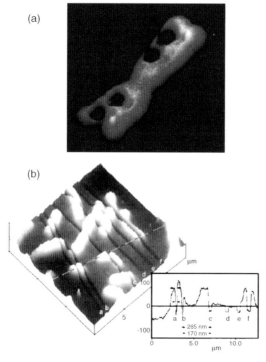

Figure 4 Femtosecond laser nanodissection of chromosomes (see Figure 2.1 and related text).

means of a photochemical effect that creates breaks in structures or pores in the membrane (Figure 4).

This mode of operation opened new kind of powerful applications, such as nano-operation of stem cells which are very useful for transfection protocols.

Dr. Gereon Hüttmann has worked for many years in the field of multiphoton imaging, mostly focusing on tissue imaging for biomedical applications. Particular attention was also devoted in both his group and that of Professor Alfred Vogel, from the same institute, to the study of photochemical effects due to the interaction of laser light with molecules. These effects form the basis of many phenomena described in this book.

Professor Daniel J. Jay has developed a method for chromophore-assisted laser inactivation (CALI) of protein function. His research is focused on proteomics, cell motility, and cancer metastasis, and he is a noted authority on protein inactivation strategies. His contribution to understanding the basis of laser–molecule interaction during optical processing of cell has been essential in revealing the range of applications also in experiments where not only breakage of structures but also selected protein inactivation was involved. This opens the way to many interesting experiments in the optical manipulation of biological processes.

Professor Andrew Woolley has carried out extensive work on photoswitching in recent years. In particular, he has worked on the photon control of peptide conformation, photoswitchable crosslinker, photososwitching photocontrolling peptide area helices, and photocontrol of protein conformation and activity. These experiments allowed, in a complementary way with respect to Daniel J. Jay, the application to other kinds of optical intervention in cell physiology, where the protein was not inactivated while its function was altered.

Professor Claus Peter Richter is one of the most well-known experts in the field of laser optical stimulation of nerves. He has worked in recent years on high-resolution mapping of the cochlear nerve using optical stimulation, revealing optical stimulation as a novel principle for neural interfaces. He has demonstrated applications also in medicine, such as the stimulation of the auditory nerve.

Professor Valery Tuchin is a recognized expert in biophotonics, biomedical optics, and laser medicine. He has authored many publications in the field of the physics of optical and laser measurements. Many interesting studies have been performed by Professor Tuchin on the diffusion in tissue of different substances and nanoparticles. In particular, he has developed an extensive imaging methodology with optical clearing agents, obtaining cell imaging in tissues with higher penetration depth and contrast.

Professor Stefan Andersson Engels' group has carried out research within biomedical and pharmaceutical laser spectroscopy applications, with many studies devoted to light propagation in turbid media. In particular, he has studied the properties of photodynamic therapy (PDT), especially with respect to different photosynthesizers. He has applied this research to oncology, with particular attention to the development of biomedical imaging applications devoted to both diagnosis and therapy of tumors.

Dr. Roberto Pini is an expert in the study of light propagation in biological tissues, including laser–tissue interaction and light excitation of organic chromophores for diagnostic and therapeutic purposes. He has also studied the synthesis, characterization, and functionalization of metal nanoparticles for medical use. He has conducted preclinical and clinical studies on the use of lasers and other optoelectronics devices in minimally invasive diagnostics and therapy with applications in ophthalmology, microvascular surgery, neurosurgery, and dermatology.

It is worth noting the development of technology in this field produced by industry. Many advances in technology have been made in the field of microscopy, with new tools for micro-dissection, catapulting, or even nanosurgery for cell transfection.

In parallel, strong chemical advances have been made and new molecular labels have been developed both for tagging purposes and photochemical manipulation of molecules.

Industry has also made notable advances in the field of light sources. New more powerful laser sources or incoherent light sources, like LEDs, have opened up new perspectives for applications. Particular attention to technological development, with respect to laser sources, has been devoted to the emission wavelength, pulse-width duration, repetition rate, and light power.

Regarding the structure of the book, Part 1 introduces the principles of multiphoton imaging, discussing two-photon microscopy and multiphoton tomography. Particular attention is devoted to endogenous cellular fluorophores and secondharmonic generation active biomolecule structures. This is followed by the principles and mechanism of femtosecond laser nanoprocessing. Such methodologies will be applied to living cells, in particular stem cells, illustrating interesting applications for transfection methodologies.

A description of nanosurgery operation to other living cells and in particular to living animals follows, opening up interesting possibilities for applications in different fields, such as neuroscience.

In Part 2, we investigate light–molecule interaction mechanisms by means of basic photophysics. Particular applications are described, such as chromophore-assisted laser inactivation, photoswitches, and optical activation of neurons.

In Part 3, we move from single cell to tissue by investigating some mechanisms of light–tissue interaction, with particular attention to tissue clearing operation. This kind of bleaching process enables an increase in penetration depth and contrast in imaging the tissue.

Continuing with the theme of tissues, in Part 4 we move from imaging to manipulation methods performed by illumination. In particular, photodynamic therapy is analyzed, starting from its cell biology base. This argument is connected to previous studies at the single cell level, described in the chromophore-assisted light inactivation operation.

Regarding PDT, immunological aspects, photophysical properties, and clinical applications are described.

Finally, and still in Part 4, another light operation on tissues modality is explained: laser welding of tissues, which is depicted with particular attention to the mechanisms of laser welding operation, together with a description of different modalities of operation and examples of clinical use.

In all parts, the cell biology aspect is described and introduced in different environments and contexts. We wish to describe both imaging and manipulation in a cell biology context, demonstrating how the light interaction and operation on complex systems such as tissues can be explained by mechanisms that rely on the cell biology base.

References

1 Denk, W., Strikler, J.H., and Webb, W.W. (1990) *Science*, **248** (4951), 73–76.

2 Goppert-Mayer, M. (1931) Uber elementarekte mit zwei quantensprunger. *Ann. Phys.*, **9**, 273.

3 Cheng, J.-X., Jia, Y.K., Zheng, G., and X. S. (2002) *Biophys. J.*, **83**, 502.

4 Freudiger, Christian W., Min, Wei, Saar, Brian G., Lu, Sijia, Holtom, Gary R., He, Chengwei, Tsai, Jason C., Kang, Jing X., and Xie, X. Sunney (2008) *Science*, **322**, 1857–1861.

5 Hell, S.W. and Wichmann, J. (1994) *Opt. Lett.*, **19**, 780–782.

6 Yanik, M.F., Cinar, H., Cinar, H.N., Chisholm, A.D., Jin, Y.,and Ben-Yakar, A. (2004) *Nature*, **432**, 822.

7 König, K., Riemann, I., Fischer, P., and Halbhuber, K.J. (1999) *Cell Mol. Biol.*, **45** (2), 195–201.

Part One
Multiphoton Imaging and Nanoprocessing

1
Multiphoton Imaging and Nanoprocessing of Human Stem Cells

Karsten König and Aisada Uchugonova

1.1
Introduction

Two-photon microscopy and multiphoton tomography with near-infrared (NIR) femtosecond lasers has revolutionized high-resolution live cell imaging [1].

Marker-free, non-destructive long-term monitoring of cells and tissues under native physiological conditions became possible. Nowadays, optical biopsies provide even better images than sliced and fixed physical biopsies [2].

Interestingly, femtosecond laser devices operating at up to three orders higher transient laser intensities than in two-photon microscopes can be used as highly precise nanoprocessing tools without collateral effects. This enables optical cleaning of cell clusters and targeted transfection of plant cells, animal cells, and human cells.

This chapter focuses on the usage of multiphoton technology for the investigation of human stem cells, one of the most interesting objects of cell biology, developmental biology, nanobiotechnology, and modern medicine.

The Russian histologist Maximow predicted the existence of stem cells 100 years ago [3]. In the 1950s, stem cells in mouse bone marrow were discovered [4]. Stem cell therapy was first demonstrated on patients with leukemia by the Nobel Prize winner Thomas at MIT in 1956. Nowadays, hematopoietic and bone marrow stem cell transplantation have become the standard therapy to treat patients with leukemia and lymphoma in combination with chemo- and ionizing radiation-therapy. There is a hope that within the next few years stem cells can be used to treat Parkinson's, Alzheimer's, cancer, diabetes, and heart diseases. In addition, stem cells will be employed to engineer tissues and to synthesize novel pharmaceutical components.

Stem cells can be classified into embryonic stem cells (ESCs) and tissue specific/adult stem cells. ESCs were isolated from mouse in 1981 [5, 6] and from humans in 1998 [7]. In 2009, the world's first human clinical trial of ESC-based therapy was approved by the American Food and Drug Administration on patients with spinal cord injuries (Geron Corp., Menlo Park, CA, USA; www.geron.com).

Laser Imaging and Manipulation in Cell Biology. Edited by Francesco S. Pavone
Copyright © 2010 WILEY-VCH Verlag GmbH & Co. KGaA, Weinheim
ISBN: 978-3-527-40929-7

So far, stem cells have to be characterized and sorted by methods that require exogenous probes and that are often destructive in 1984. It is of great interest to develop non-destructive, marker-free *in vivo* techniques to detect and to manipulate stem cells.

Multiphoton imaging is the ideal technique to trace and image the stem cells over a long period of time as well as to study their differentiation process without any marker. Furthermore, multiphoton technology can be used for optical cleaning and optical DNA transfer into stem cells. This chapter focuses on multiphoton technology for human stem cell research.

1.2
Principle of Two-Photon Microscopy and Multiphoton Tomography

Confocal one-photon (linear) microscopy was invented by Marvin Minsky in 1957. With the availability of lasers, beam scanning systems, and sensitive photomultipliers (PMT), the first commercial confocal laser scanning microscopes (CLSMs) for 3D imaging became available in the 1980s. ZEISS Jena and the University of Jena built a confocal picosecond laser scanning microscope with temporal and spatial resolution in 1988 [8–10]. Thousands of CLSMs are currently operating in the field of life sciences, mainly based on one-photon excited fluorescence where the visible (VIS) intracellular fluorescence of exogenous molecular and cellular probes is detected with submicron resolution. Optical sectioning enables the 3D reconstruction of the object of interest.

Conventional one-photon fluorescence microscopes employ UV and VIS light sources, such as the argon-ion laser at 364/488/515 nm emission, the frequency-converted Nd:YAG laser at 355/532 nm, and the helium neon laser at 543/633 nm. Typically, the fluorescence excitation power on the target is some microwatts, which corresponds to light intensities in the range of $kW\,cm^{-2}$ when focused to diffraction-limited spots by objectives with high numerical aperture (NA > 1).

Wilson and Sheppart proposed the application of nonlinear excitation to microscopy in 1984. Denk, Strickler, and Webb realized, finally, the first two-photon NIR microscope based on a femtosecond dye laser in 1990 [1]. Multiphoton absorption requires high light intensities in the range $100\,MW\,cm^{-2}$ up to $TW\,cm^{-2}$. In principle, continuous wave (CW) NIR radiation can induce two-photon effects if the power exceeds 100 mW and when high NA objectives are used [11–13]. To avoid trapping effects and to reduce the mean power, most multiphoton microscopes are based on femtosecond laser pulses at MHz repetition rate with high kilowatt peak power (P) and low mean power in the $\mu W/mW$ range. The multiphoton efficiency of an n-photon process follows a P^n relation. In the case of two-photon microscopy, the two-photon effect depends on P^2/τ. The shorter the pulse width (τ) and the higher the laser power the more fluorescence photons and the better the nanoprocessing. Multiphoton microscopy at a fast scanning rate with microsecond beam dwell times per pixel is possible. Figure 1.1 demonstrates the principle of two-photon microscopy where two NIR photons are absorbed simultaneously to induce a single VIS fluorescence photon.

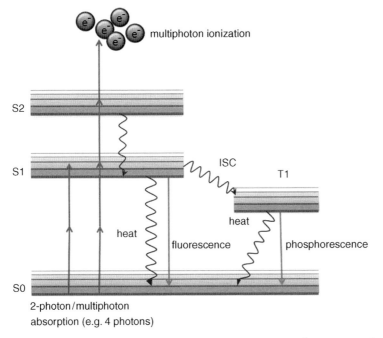

multiphoton ionization

S2

S1

ISC

T1

heat

heat

fluorescence

phosphorescence

S0

2-photon/multiphoton
absorption (e.g. 4 photons)

Figure 1.1 Principle of two-photon excited fluorescence at GW cm^{-2} intensities and principle of multiphoton ionization, which requires multiple photons (e.g., 4 or 5) and TW cm^{-2} intensities.

Multiphoton microscopes, including two-photon and three-photon fluorescence microscopes, second-harmonic generation (SHG) and third-harmonic generation (THG) microscopes as well as nanoprocessing microscopes, are based on the application of low energy photons in the NIR between 700 and 1200 nm. This spectral range is also referred as an "optical window" where the one-photon absorption coefficients and scattering coefficients of unstained cells and tissues are low. Most cells (except erythrocytes and melanocytes) appear transparent. The light penetration depth in tissue in this spectral region is high and in the range of some millimeters.

A significant advantage of multiphoton high NA microscopy compared to conventional one-photon microscopy is the tiny sub-femtoliter excitation volume. Absorption in out-of-focus regions is avoided because the probability of two-photon absorption decreases nearly with the distance d from the focal point according to a d^{-4} relation.

In particular, when studying 3D objects, including cell clusters, embryos, and tissues by optical sectioning, multiphoton microscopy is the superior method compared to one-photon confocal scanning microscopy with its large excitation cones and the subsequent problem of out-of-focus damage as well as the disadvantage of using high-energy excitation photons.

Long-term studies have demonstrated that multiphoton microscopy can be performed without photodamage under certain conditions. In particular, single hamster ovarian cells have been femtosecond laser-exposed for hours with a high 200 $GW\,cm^{-2}$ peak intensity without any impact on cellular reproduction and vitality [14]. In another long-term study, living hamster embryos were exposed for 24 h with a multiphoton NIR microscope and implanted in the mother animal without impact on embryo development in contrast to control studies performed with a conventional one-photon CW VIS laser microscope [15].

The first commercial multiphoton tomograph, DermaInspect™ (JenLab GmbH), is in clinical use for melanoma diagnosis and *in situ* intradermal drug targeting. It was shown that this femtosecond laser system is safer than conventional UV light sources used in the cosmetic industry and in tanning studios [16].

In addition to conventional one-photon microscopes, multiphoton microscopes enable SHG and THG imaging. In SHG, two photons interact simultaneously with non-centrosymmetrical structures such as collagen and generate coherent radiation at exactly half of the excitation wavelength in forward direction. There is no light absorption. Therefore, photobleaching and photodamage are excluded. SHG enables deep 3D imaging due to backscattered light.

Multiphoton microscopes do not require confocal units or de-scanned detection systems due to the sub-femtoliter excitation volume.

1.3
Multiphoton Microscopes and Multiphoton Tomographs

Owing to patent and marketing issues, most commercial two-photon microscopes are based on expensive confocal microscopes with the extension of a NIR femtosecond laser combined with a special interface. This makes the two-photon microscopes of the major microscope suppliers very expensive and avoids the sale of more compact pinhole-free two-photon microscopes. With the end of patent rights in 2009/2010, low-price ultracompact two-photon systems will become available.

Figure 1.2 shows two-photon microscopes. The first is the ZEISS META-LSM510-NLO confocal microscope (Figure 1.2a). The second is the pinhole-free ultracompact laser scanning microscope "Two-photon TauMap" where the galvoscanners and the photon detectors were attached to the side ports of an inverted microscope (JenLab GmbH, Jena, Germany).

In most microscope systems, compact solid-state "turn-key" tunable 80/90 MHz titanium:sapphire lasers such as the "Chameleon" from Coherent, USA, and "MaiTai" from Newport/Spectra Physics, USA, are employed. The laser beam is typically transferred to a beam expander, a beam attenuator, the scanning optics, x/y-galvoscanner, and focused into the target by a piezo-driven objective of high numerical aperture (NA > 1). The signal is typically measured by a sensitive photomultiplier (PMT).

Exogenous fluorescent probes with high fluorescence quantum yields (e.g., DAPI and Hoechst) require a mean power of 25 to 100 µW at a frame rate of 1 Hz (512 × 512 pixels). The endogenous intracellular fluorophore NAD(P)H can be imaged with less

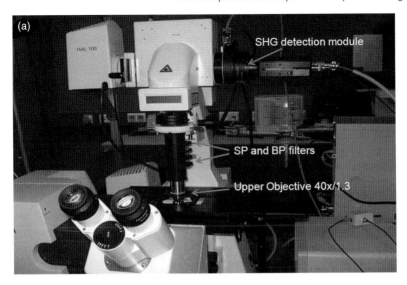

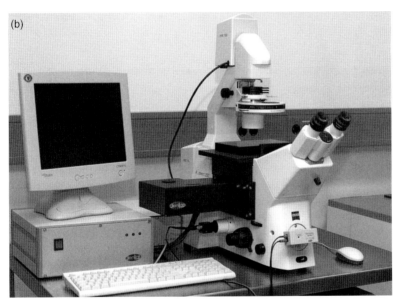

Figure 1.2 Photographs of a ZEISS META LSM510-NLO system with an additional SHG detection module (a) and a compact Two-photon TauMap Microscope (b).

than 2 mW mean power at appropriate NIR wavelengths with typical exposure times of 1 to 8 s per frame.

Optical dispersion results in pulse broadening during transmission through the microscope. Typically, the pulse width at the sample is about 150–300 fs.

Recently, the ultracompact nanoprocessing microscope FemtOgene™ (JenLab GmbH, Jena, Germany) with the shortest femtosecond laser pulses at the target of 12 femtoseconds was released onto the market. The ultrashort pulse width was achieved by dispersion pre-compensation of the whole microscope, including the beam expander, the polarizer for beam attenuation, the tube lens, the objective, filters, and so on, using chirped mirror technology (Figure 1.3).

The microscope can be used in the two-photon fluorescence excitation mode at mean powers in the microwatt range for nondestructive imaging of the stem cell of interest and to monitor the biosynthesis of fluorescent proteins. Furthermore, it can be employed for nanoprocessing when operating in the milliwatt power range in three exposure modes: (i) scanning of a region of interest (ROI) for ablation, (ii) line scanning for cutting, and (iii) single point illumination for drilling where the galvoscanners are fixed to a point of interest.

Nowadays, femtosecond laser tomographs for high-resolution deep-tissue imaging in animals and humans are operating as CE-marked medical devices in Europe, Asia, and Australia. Figure 1.4 shows the two types of multiphoton tomographs DermaInspect and MPTflex™ that are currently available on the market. They are employed for small animal studies based on the detection of fluorescent proteins as well as tissue engineering and for clinical studies regarding skin cancer diagnosis, wound healing studies, and nanoparticle tracking.

1.4
Endogenous Cellular Fluorophores and SHG Active Biomolecule Structures

Non-invasive multiphoton NIR microscopes [1, 14, 17, 18] have been applied to image living single cells and different tissues with a high spatial resolution without any staining. Two-photon autofluorescence can be obtained from intrinsic fluorophores such as NAD(P)H, flavins, porphyrins, elastin, and melanin [19]. In addition, SHG images can be obtained from certain biomolecule structures such as collagen and myosin [15, 20–24].

The most important endogenous cellular fluorophores for two-photon imaging are the reduced coenzymes nicotinamide adenine dinucleotide (NADH) and nicotinamide adenine dinucleotide phosphate (NADPH), referred as NAD(P)H, with a broad emission in the blue/green spectral range. The oxidized form NAD(P) shows no significant VIS fluorescence. The reduced coenzymes possess a folded and an unfolded configuration, with the unfolded ones to bound NAD(P)H. The emission maximum of the unfolded configuration shows a blue-shifted maximum (450 nm) and a higher fluorescence quantum yield than free NAD(P)H with its maximum at 470 nm [19]. The hydrogen-transferring pyridine coenzymes are mainly located in the mitochondria and play a key role in respiratory chain activity and act as sensitive indicators of the cellular metabolism since the metabolic activity of cells is given by the ratios of the concentrations of free to protein-bound NAD(P)H and of NAD(P)H to flavins [19, 25–27]. Normally, the excitation of NAD(P)H requires UV light at around 340 nm. However, the use of UV exposure should be avoided due to photoinduced cytotoxic reactions and the limited light penetration depth.

(a)

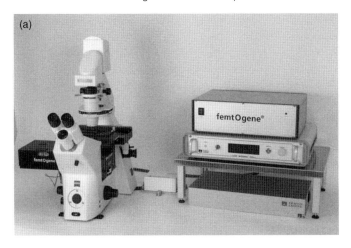

(b)

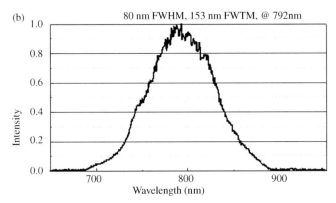

(c)

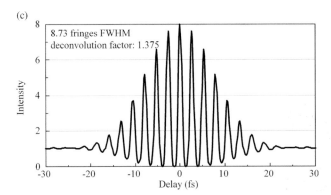

Figure 1.3 Photograph of the imaging and nanoprocessing 12-femtosecond laser microscope Femtogene (a) as well as a spectrum (b) and autocorrelation function at the focus of the 40 ×, NA1.3 objective (c).

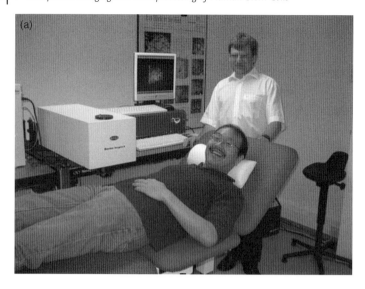

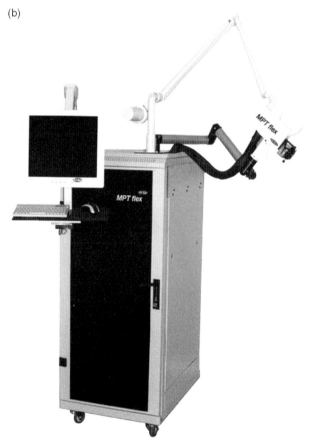

Figure 1.4 Photographs of the multiphoton tomographs DermaInspect (a) and MPTflex (b).

Two-photon NIR excitation is the ideal method to image intracellular/intratissue NAD(P)H without the disadvantages of UV microscopy.

Cellular flavins such as flavin adenine dinucleotide (FAD), flavin mononucleotide (FMN), lipoamide dehydrogenase (LipDH), riboflavin, and electron transfer flavoprotein (ETF) are also fluorescent coenzymes that are involved in oxidation---reduction reactions [26]. The excitation spectra of these flavins/flavoproteins possess major electronic transitions at 260, 370, and 450 nm as well as an emission peak around 530 nm. The fluorescence intensity decreases when covalently attached to proteins ("flavoprotein," [19]).

Besides the measurement of the fluorescence/SHG/THG intensity by optical sectioning (3D imaging), fluorescence lifetime imaging (FLIM, 4D) and spectral imaging (5D) can be performed. In particular, the arrival times of the fluorescence photons with respect to the excitation time of the molecule and the particular location (pixel) can be determined by time-correlated single photon counting (TCSPC) and the use of photomultipliers with short rise time. When using a PMT array in combination with a polychromator, the "color" of the emitted photon per pixel can be also determined (spectral imaging). Recent developments involve PMT arrays with fast picosecond rise time in combination with a polychromator or other wavelength-selective components to realize spectral FLIM within one scan [28].

1.5
Optical Nanoprocessing

1.5.1
Principle and Mechanism of Femtosecond Laser Nanoprocessing

Femtosecond laser microscopes can be used as precise nanoscissors based on multiphoton ionization at TW cm^{-2} intensities. The simultaneous absorption of several photons, for example, five photons, induces ionization of molecules and the generation of free electrons. The onset of plasma occurs if a density of 10^{21} electrons cm^{-3} is achieved [29]. In the case of water, a "bandgap" of 6.5 eV has to overcome. Since 800-nm photons have a photon energy of 1.55 eV, five photons are required for multiphoton ionization of water molecules [29]. The application of 750-nm photons (1.65 eV) requires four photons. The formation of plasma in aqueous solution results in the formation of plasma-filled cavitation bubbles that can be video-imaged. Interestingly, nanoprocessing of cells can be performed at thresholds below the threshold for multiphoton ionization of water.

The bubbles expand and can induce an implosion combined with destructive jet streams and shock waves. These photodisruptive effects result in destruction of the microenvironment. The onset of photodisruptive effects depends on the transient laser intensity whereas the amount of damage depends on the pulse energy. The smaller the applied pulse energy the smaller the photodisruptive effects [30]. Therefore, low pulse energies are required to perform nanoprocessing without collateral destructive effects. The procedure should, consequently, be performed near the threshold of multiphoton ionization and plasma formation of cellular structures. This can be realized by fine-tuning of the laser power. In early studies on chromosome cutting, pulse energies of about 1 nJ were applied when using 200–300-femtosecond laser pulses at 80 MHz repetition rate [31]. Precise laser nanosurgery depends on laser wavelength, laser intensity, pulse duration, repetition rate, and irradiation time.

The first femtosecond laser subcellular nanosurgery was demonstrated by König's group in 1999 by dissections of single chromosomes in living cells [31–34]. The group also introduced femtosecond laser assisted single cell transfection by creating transient holes in the cellular membrane [35]. Femtosecond laser pulses have been applied to dissect single dendrites and cytoskeletons, to optically knock out intracellular mitochondria [36–38], for membrane and microtubules surgery [39], and for ocular refractive surgery [40].

1.5.2
Stem Cells

Adult stem cells, also termed somatic or tissue specific stem cells, can be found among differentiated cells within fully developed tissues in a low abundance. For example, mesenchymal stem cells represent 0.01–0.0001% of the total population of nucleated cells in the bone marrow [41, 42]. Mesenchymal stem cells have been also isolated from bone, skin, muscle, adipose, cartilage, peripheral, and umbilical cord blood, whereas hematopoietic stem cells have been isolated from peripheral and umbilical cord blood as well as bone marrow. Neural stem cells have been isolated from different parts of the brain. Stem cells were found in children's primary teeth, hairs, retina, skin, lung, liver, and glands.

Adult stem cells are "multipotent" and differentiate mainly into several specialized cell types depending on the kind of tissue they have been originated from. However, several studies have also shown that cells originated from the mesoderm (muscle, blood, bone, fat) can produce cells normally originated from the endoderm (gut, lungs, liver, and pancreas) and the ectoderm (skin, nervous system) germ layer. Under laboratory conditions, directed differentiation can be induced by the stimulation of certain factors such as the growth factor TGF-β and cytokines. Nanostructures, mechanical stress and vibration, temperature, light, and electrical fields can also induce differentiation [43–46]. The proliferation and differentiation capacity of adult stem cells is limited because they stop dividing after several passages due to a decrease of the telomerase activity and telomere shortening [47–49]. Furthermore, a prolonged culture of adult stem cells can induce tumorogenity [50]. Freshly isolated adult stem cells do not show a propensity for tumor formation. Undesired differentiation occurred under laboratory conditions [51].

More than 2000 patients with Parkinson's disease, Alzheimer's disease, cerebral palsy, diabetes mellitus, spinal cord injury, and cardiac infarction have been successfully treated with autologous adult stem cells from bone marrow in Germany (X-Cell Center at the Institute of Regenerative Medicine, Germany; www.xcell-center.de).

No side effects have been reported. However, the exact repair phenomenon remains to be understood. It would be helpful to trace individual stem cells in culture as well as *in vivo* to study their interaction with the microenvironment over a long period of time. Unfortunately, so far it is difficult to identify, to isolate, and to purify adult stem cells from the tissues.

Embryonic stem cells can be grown in their non-differentiated state for many generations. They can produce cells from the three germ layers, including germ cells under laboratory conditions. The high potential of ESCs to treat diseases has been tested on animal models successfully. Fewer studies have been conducted on human ESCs because of ethical constraints. A tumorogenity of ESCs has been reported upon transplantation [52]. Therefore it is recommended to differentiate cells before transplantation.

Tissue engineering from the patient's own cells is a further major application field that would overcome the problems of the transplanted purely artificial and mechanical prosthesis that cannot display physiological function and missing self-repair. So far various tissues/organs like bladder, blood vessels, skin, cartilage, heart valve, kidney, liver, salivary glands, pancreas, ear, and bone have been generated by tissue engineering and clinical trials have been conducted [53].

Stem cells can be genetically modified for stem cell therapy and the development of specific cell populations. Owing to the restriction of the isolation of ESCs from blastocytes, there is an interest in producing ESC-like cells from somatic cells through reprogramming and nuclear transfer. In 2007, it was demonstrated that induced pluripotent stem cells can be generated from adult dermal fibroblasts by introducing the four genes Oct3/4, Sox2, Klf4, and c-Myc [54].

To find stem cells in tissues, biopsies have to be taken, sliced into microsections, and tagged with the marker. Typically, fluorophores with a high quantum yield are employed in this immunocytochemistry approach where the stem cells are no longer alive. The genetic approach requires procedures to gain, to amplify, and to stain DNA that also destroy the stem cells.

Fluorescence-assisted cell sorting (FACS, a special type of flow cytometry) can be applied to identify and isolate living stem cells in suspension. Hundreds of thousands of cells marked within a fast jet stream pass a laser beam that induces fluorescence from the cells only that are tagged with specific fluorescent stem cell markers. When fluorescence occurs (selection criteria), an electromagnetic field is switched on that charges the fluorescent cell. This particular cell within a droplet can be deflected by a strong electrostatic field and can be collected. FACS enables precise counting and accurate sorting of stem cells of interest. The major disadvantages of FACS are the low viability of sorted cells.

In view of the various disadvantages of the different identification and sorting techniques, it would be of great importance to find chemical-free methods to identify stem cells in their native 3D tissue environment as well as in *in vitro* cell suspensions and biopsies. It would also be of tremendous interest if a marker-free imaging method could be employed to monitor stem cells and their differentiation over a long period of time under physiological conditions.

1.5.3
Upgrading the Multiphoton Microscope

Three-dimensional multiphoton microscopy (x,y,z) is based on optical sections in different tissue depths. One section is obtained by beam scanning using a fast x–y galvoscanner in a typical 512×512 pixel field covering an area of $230 \times 230\,\mu m^2$ at $25.6\,\mu s$ pixel dwell time. To change the focal plane either the motorized stage can be moved in the axial (z) direction (ZEISS LSM510-NLO) or the focusing optics is moved by a piezosystem (FemtOgene, Two-photon TauMap, MIPOS 5, $500\,\mu m$ working distance, Piezosystems Jena, Germany). 3D images are obtained by the correlation of the PMT signal with the position of the x–y galvoscanners and the z-position of the stage/focusing optics. The microscope can be upgraded to a fourth dimension system (4D microscopy: x,y,z,τ), which depicts the fluorescence lifetime as false color into the high-resolution image.

Upon photoexcitation, within femtoseconds from the ground state (S_0) to a higher electronic state (S_1), the molecule will remain in the excited state only transiently for some picoseconds up to tens of nanoseconds. This average time a fluorescent molecule remains in the excited state is referred to as the "fluorescence lifetime, τ." The parameter τ is a signature of the fluorescent material and independent of the concentration, illumination intensity, light path of the optical system, and detector [55]. The fluorescence lifetime depends on the microenvironment and changes as a result of the interaction with other molecules due to the loss of their excited state energy by additional decay pathways.

Since its introduction to life sciences 20 years ago [9, 10], fluorescence lifetime imaging microscopy (FLIM) has become a key technique for imaging cellular processes, protein–protein interactions, and tissue compartments [56, 57].

The fluorescence lifetime is often determined by time-correlated single photon counting (TCSPC) where the arrival times of photons are measured with respect to the excitation pulse. Thousands of photons are counted per pixel and placed into different "time channels" to build up a histogram and a fluorescence decay curve, $F(t) = F_0 e^{-t/\tau}$, respectively. The fluorescence lifetime can be calculated from this decay curve with an accuracy according to Poisson statistics [55]. Often, the fluorescence decay curve represents a multi-exponential decay due to the presence of different fluorescent molecules in one "pixel" or the presence of one fluorophore in its free and its bound form. Typical fitting procedures can consider a mono-exponential as well as a bi-exponential behavior, $F(t) = A_1 e^{-t/\tau_1} + A_2 e^{-t/\tau_2}$.

Sometimes, the intensities and the fluorescence lifetimes of two different fluorophores are similar. It would be helpful to have a further criterion to distinguish between them. This can be the "color" of the emitted photon, which can be determined by separation of the photons into "spectral channels." The method is also called "emission fingerprinting" [58]. The combination with 4D microscopy enables "spectral FLIM" (5D microscopy: x,y,z,τ,λ). Spectral imaging can be performed, for example, with a 32-channel PMT array (ZEISS-META, Hamamatsu) in combination with a polychromator with a resolution of 10.5 nm per channel. A full 512×512 lambda stack of data from all 32 channels (the full visible spectrum

382–714 nm) can be acquired in 0.8 s. In addition, a further fast PMT can be used for FLIM. More elegant is the employment of an array with fast PMTs in combination with a polychromator as well as a multichannel FLIM module such as the SPC830 from Becker & Hickl GmbH, which collects the photons according to the arrival time as well as their color.

1.5.4
Autofluorescence Imaging of Human Stem Cells

Two-photon excitation of endogenous fluorophores enables non-destructive high resolution imaging of living cells and extracellular matrix (ECM) components over a long period of time. The application of NIR femtosecond laser pulses induced a blue/ green cellular autofluorescence that provided information on the cell morphology and cell size as well as enabled visualization of some cell structures and ECM components with submicron resolution without exogenous markers. The most intense fluorescent structures were found to be the mitochondria. Table 1.1 shows typical laser and exposure parameters for safe stem cell imaging with either the ZEISS LSM510-NLO microscope with 250 fs pulse width at the target and a preferred excitation wavelength of 750 nm or the ultracompact FemtOgene 12-fs microscope. Figure 1.5 demonstrates autofluorescence images of human salivary gland stem cells (hSGSCs), human dental pulp stem cells (hDPSCs), and human pancreatic stem cells (hPSC) acquired at 750 nm excitation wavelength. The ellipsoidal/round non-fluorescent nucleus and the fluorescent mitochondrial network are clearly seen.

When using 750 and 800 nm, NAD(P)H as well flavins/flavoproteins have been efficiently excited by a two-photon process. When changing to 850 nm or even 900 nm, flavins/flavoproteins (e.g., FAD) only and not NAD(P)H were efficiently excited.

Spectral measurements showed a fluorescence maximum at 460–470 nm when excited with 750 nm light, which is consistent with the emission behavior of free and protein bound NADH. The maximum shifted to 530–535 nm when using 900 nm light, which is consistent with flavin emission (Figure 1.6).

Spatially resolved autofluorescence decay curves were obtained at 750 and 900 nm and a scanning time of up to 30 s per frame. Figure 1.7 represents a typical FLIM image, a histogram, and a particular intramitochondrial fluorescence decay curve of 750 nm-excited hSGSC stem cells based on 6740 detected photons. The bi-exponential fit with the optimal fitting parameter $\chi^2 = 1.00$ revealed a fast decaying fluorophore with a short lifetime τ_1 of 0.17 ns and an amplitude $a_1 = 72\%$ and a second component with $\tau_2 = 1.8$ ns and $a_2 = 28\%$. Although the amplitude is lower, the longer component provides $\tau_2 a_2 / \tau_1 a_1 = 4$ times more fluorescence intensity than the short-lived fluorophore.

1.5.5
Multiphoton Imaging during Differentiation

Human SGSCs stem cells can undergo adipogenic differentiation resulting in the formation of mature adipocytes.

Table 1.1 Typical laser parameters and exposure parameters used for two-photon imaging of stem cells.

Pulse width, τ (fs)	Mean power, P (mW)	Wave length, λ (nm)	Spot size, d (μm)	Repetition frequency, f (MHz)	Pulse energy, E (pJ)	Peak power, P_{max} (kW)	Mean light intensity, I (MW cm^{-2})	Peak light intensity, I_{peak} (GW cm^{-2})	Exposure time, t (μs)	Energy density (J cm^{-2})	Laser pulse number, n	Applied energy (nJ)
250 (LSM 510)	5	750	0.58 (NA 1.3)	80	62.5	0.25	1.9	95	6	11.4	480	30
12 FemtOgene	0.5	800	0.62 (NA 1.3)	75	6.6	0.55	0.2	182	6	1.2	450	3

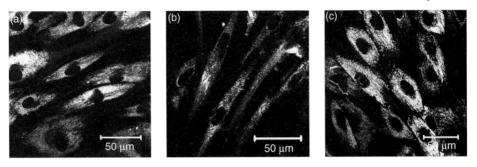

Figure 1.5 Two-photon autofluorescence images of human stem cells excited at 750 nm. (a) Human salivary gland stem cells, (b) human dental pulp stem cells, and (c) human pancreatic stem cells [59].

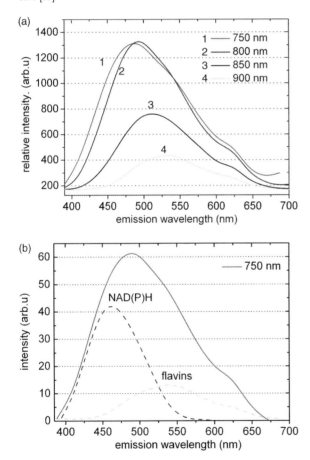

Figure 1.6 Spectral measurement of hSGSC stem cells. (a) Four excitation wavelengths were employed. The emission peak shows a redshift with increasing excitation wavelength due to the preferred two-photon excitation of flavins versus NAD(P)H. (b) The 750/800 nm excited emission spectra can be considered as an overlay of the two fluorophores NAD(P)H and flavins (dotted lines, spectral unmixing).

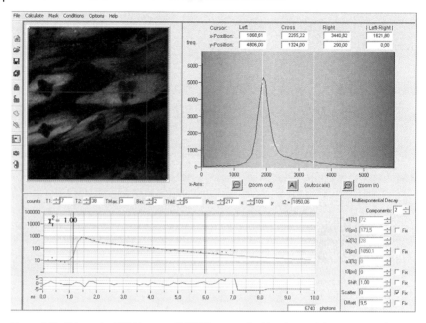

Figure 1.7 False-color FLIM image of hSGSC stem cells. The depicted decay curve is from an intracellular pixel of the right cell (cross), the τ_2-histogram from the whole frame [59].

Figure 1.8 shows images of about ten stem cells inside a spheroid originated from hSGSC stem cells taken one week after administration of adipogenic differentiation factors. Two cells are differentiated into about 100-μm long adipocytes with mainly round, non-fluorescent lipid vacuoles with a diameter up to 10 μm that store fat. When analyzing the gray levels of the autofluorescence pattern, the differentiated cells were found to possess lower fluorescence intensity than cells without fat droplets. Spectral imaging at 750 nm excitation wavelength revealed significant changes of the emission spectrum. Non-differentiated highly-fluorescent cells exhibited a maximum at 490 nm with shoulders at 540 and 620 nm, whereas the adipocytes emitted at 460 nm with a less pronounced shoulder at 620 nm. The maximum at 460 nm corresponds to the typical emission maximum of NAD(P)H. FLIM data from highly fluorescent vacuole-free stem cells revealed a τ_m of 0.59 ns, a short 0.22 ns decay component (78%), and a long 1.93 ns decay component. In contrast, the mean lifetime of cells with fat droplets was much longer with typical values of more than 0.85 ns.

In addition to two-photon excited fluorescence, two-photon microscopy allows the detection of SHG in particular from the ECM component collagen. Collagen also shows weak autofluorescence. When using excitation wavelengths shorter than 800 nm, SHG cannot be detected due to technical reasons (UV absorbing filters). The 750-nm-excited autofluorescence decays reveal a short component of about 0.26 ns lifetime with an amplitude of 58% whereas the longer component has a lifetime of 2.6 ns (42%). By contrast, the signal is mainly based on SHG (98%) when

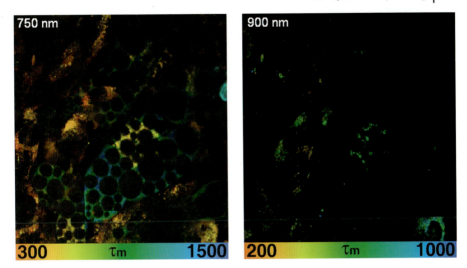

Figure 1.8 False-color coded τ_m FLIM images of cells excited at 750 and 900 nm, respectively, after adipogenic differentiation. At the higher excitation wavelength, no NAD(P)H fluorescence is observed.

using an excitation wavelength of 850–900 nm. Therefore, false-color coded FLIM images can be used to differentiate between SHG and autofluorescence (Figure 1.9). The biosynthesis of collagen in 3D salivary gland and pancreatic stem/progenitor spheroids has been monitored after incubation in the chondrogenic differentiation medium. The differentiation process has been observed for a long time period (up to 5 weeks). The first SHG signal was detected eight days after the introduction of the stimulating agents.

In addition, SHG radiation was detected in the case of osteogenic differentiation from pancreatic, salivary gland, and bone marrow mesenchymal stem cells.

1.5.6
Nanoprocessing

The sub-20 fs microscope FemtOgene, as well as the 250 fs ZEISS laser scanning microscopes, can perform nanoprocessing by single point illumination and line scanning at mW mean powers. Sub-20 fs laser pulses required significant less power than the long 250 fs laser pulses. In fact, very precise cuts within the cytoplasm and the nucleus could be performed without any collateral damage at very low average powers, as low as 7 mW, when using 12 fs laser pulses. Video-taping revealed the formation of some small microbubbles with a lifetime of less than 2 s. When increasing the power up to 20 mW, more destructive effects occurred, due to the occurrence of several bubbles with sizes up to 5 μm. The most efficient way to destroy a single cell of interest was found by scanning the whole cell for a few seconds.

(a)

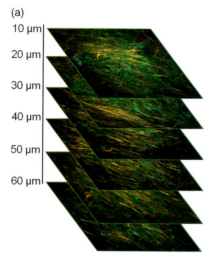

(b)

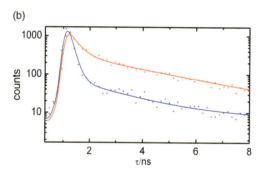

Figure 1.9 (a) Multiphoton sections of stem cell spheroids after chondrogenic differentiation (SHG: yellow fibrils; autofluorescence: blue/green). (b) luminescence decay curves showing the autofluorescence decay of a single fluorescent organelle (upper curve) and the SHG/autofluorescence decay from a collagen structure (lower curve).

To expose single cells inside the stem cell spheroids, at first 3D two-photon autofluorescence optical sectioning was performed. Optical destruction was carried out by single point illumination within the spheroid. In particular, one cell was exposed with high pulse energy (2.6 nJ) and an exposure time of 10 ms. The laser-exposed cell took up the dead-cell indicator ethidium bromide and appeared as very bright fluorescent single cell at the depth of 20 μm.

The method of selective optical destruction of cells of interest can be described as optical knock out [60]. The method has been used to demonstrate the possibility of "optical cleaning" of cell cultures. Stem cell cultures were cleaned in such a way that a particular cell of interest remained alive while the surrounding cells obtained a lethal laser exposure. Figure 1.10 demonstrates the procedure. To isolate a single cell of interest, the surrounding cells within a $0.7 \times 0.7 \, mm^2$ area were exposed

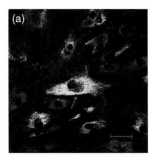
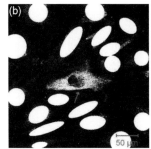
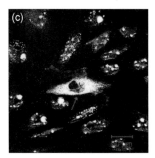

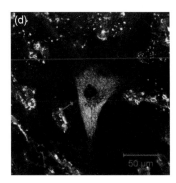

Figure 1.10 Optical cleaning of stem cell cultures. (a) Autofluorescence image before optical knock-out; (b) determination of regions of interest (white), which should be exposed with an intense laser beam at TW cm^{-2} intensity; (c) autofluorescence image taken 1 min after optical knock-out; (d) image after one day. The cells without intense laser exposure survived whereas the surrounding exposed cells were destroyed.

with 200 mW average power and ROI scanning. Significant changes of the morphology and the autofluorescence pattern of the laser exposed cells were monitored within seconds of the intense laser exposure. The non-exposed cell remained intact and was monitored over several days. A normal division of the isolated single living cell was detected as well as the migration of new cells to the laser-exposed area.

The contact-free and chemical-free nanoprocessing method was also probed to realize targeted transfection of single stem cells. For that purpose, single point illumination was performed when opening the shutter for 50 ms, which corresponds to 3.75 million pulses. Figure 1.11 shows two-photon images of an optoporated cell after introduction of the membrane-impermeable GFP plasmid pEGFP-N1 from Clontech (4.7 kb, molar weight 3 MDa) and after cell division occurred. The formation of green fluorescent proteins with an emission maximum at 507 nm was monitored. Interestingly, all laser-exposed cells survived. The green fluorescent daughter cells indicate the successful GFP biosynthesis after introduction of the plasmid into the cytoplasm through the nanopore and the uptake into the cellular DNA as well as the successful reproduction of laser-exposed cells.

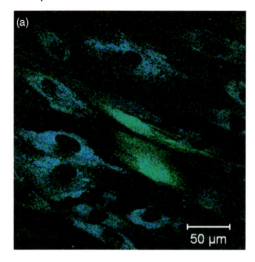 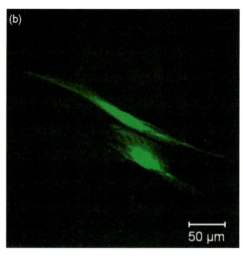

Figure 1.11 Transfection of targeted salivary gland stem cell in an adherent cell culture. A single cell was optoporated. The transfected cell successfully divided and became two-green fluorescent daughter cells. (a) NAD(P)H fluorescence (blue/green) and GFP fluorescence (green) appeared when excited at 750 nm. (b) 800 nm excitation induced mainly GFP emission.

1.6
Discussion and Conclusion

Marker-free multiphoton autofluorescence/SHG microscopy has been used to image stem cells and their development into specialized cells up to five weeks after introduction of a differentiation-stimulating medium. Significant changes of the autofluorescence occurred.

During adipogenic differentiation, the blue-shifted emission maximum and the relative increase of the long fluorescence lifetime (τ_2) indicate the increased ratio of bound to free NAD(P)H (Table 1.2). Because there is an overall decrease of the fluorescence intensity during differentiation, the concentration of the reduced form NAD(P)H drops compared to the oxidized from NAD(P) and therefore the oxygen consumption increases. This hypothesis is in accordance with published data [61] that claims a decrease of NAD(P)H/NAD$^+$ means an increased metabolism. More studies on various stem cells under different types of differentiation are required to understand the exact metabolism during the differentiation process.

In the case of osteogenic and chondrogenic differentiation, SHG can be used to detect the biosynthesis of collagen [59, 62].

As shown, sub-20 fs laser scanning microscopes can be employed for nano-injection and transfection with low mean powers of 5–7 mW (66–93 pJ @ 75 MHz), which is more than one-order less power than in current femtosecond laser

Table 1.2 Autofluorescence parameters of human salivary gland stem cells (hSGSCs) and cells after adipogenic differentiation (adipocytes) measured by 5 D multiphoton imaging.

Cells	NAD(P)H fluorescence intensity	Fluorescence maximum (nm)	Flavin fluorescence intensity	NAD(P)H/ flavin	Fluorescence lifetime (ns)	Bound NAD(P)H/ free NAD(P)H
hSGSCs	High	493	High	Low	Short 0.22/2.00	Low $a_2/a_1 = 0.28$
Adipocytes	Low	Blue-shifted 460	Low	High	Long 0.26/2.33	High $a_2/a_1 = 0.50$

nanoprocessing tools. The use of a low sub-10 mW power level avoids destructive thermal effects and trapping effects and leads to high transfection efficiencies. The potential use of such low power systems opens the way for the manufacturing of ultracompact low-power laser systems for nanoprocessing as well as for imaging at a lower price than current femtosecond laser microscopes.

Stem cells are now at the "bench to bedside" stage for the treatment of myocardial infarction, neurological diseases such as Alzheimer and Parkinson, diabetes, and cancer. Current stem cell therapy is based on adult stem cells where the recruitment of the limited number and their restricted locations is extremely difficult. Furthermore, the stem cells of the patient with genetically based diseases will also carry the genetic effect. There is a hope that multiphoton imaging and nanoprocessing tools can be employed to trace and to manipulate the stem cells as well as to monitor and to influence the differentiation process. However, future work on a large number of different stem cells within their native tissue environment has to be conducted to answer the question of whether the rare stem cells can be identified due to their characteristic autofluorescence behavior. One recent paper reports on the prospective isolation of bronchiolar stem cells from facultative transit-amplifying cells by autofluorescence detection [63]. Results from frozen tissue sections have also demonstrated a different autofluorescence behavior of stem cells in the hair follicle bulge compared to the surrounding cells [64]. Multiphoton microscopes have also been used to monitor the migration of neural stem/progenitor cells in the brain of mice [65].

Current multiphoton microscopes enable the study of stem cells *in vitro*. However, the first two-photon tomographs for skin imaging are in clinical use [2]. There is a chance to trace adult stem cells in the basal cell layer of the epidermis and along the hair shafts. In the near future, two-photon microendoscopes will be available [66] that can be employed to trace stem cells inside the body.

As demonstrated, a multiphoton system cannot only be employed as a high-contrast and high-resolution imaging system but also as highly precise nanoprocessing tool, for example, or optical cleaning and targeted transfection [60, 67, 68]. The optical generation of transient nanoholes will open up a novel way to deliver not only foreign DNA but also other different molecules and chemicals such as RNA, recombinant proteins, nanoparticles, and drugs into the living cell without destructive collateral effects. Very recently it has been demonstrated that the microscope FemtOgen can be employed to deliver super quencher molecular beacon (SQMB) probes into living cells [69].

Future engineering work may result in the development of fiber-based systems with automatic miniaturized high-throughput femtosecond laser systems, for example [70].

In conclusion, multiphoton femtosecond laser systems open up new opportunities in the investigation of human stem cells and their medical application. They can provide morphological and biochemical information with single molecule sensitivity of stem cells in their native environment and can be used to transfect, knock-out, and sort the cells.

References

1 Denk, W., Strickler, J.H., and Webb, W.W. (1990) Two-photon laser scanning microscope. *Science*, **248**, 73–76.

2 König, K. (2008) Clinical multiphoton tomography. *J. Biophoton.*, **1**, 13–23.

3 Maximow, A. (1909) Der Lymphozyt als gemeinsame Stammzelle der verschiedenen Blutelemente in der embryonalen Entwicklung und im postfetalen Leber der Säugetiere. *Folia Haematol.*, **8**, 125–141.

4 Siminovitch, L., McCulloch, E.A., and Till, J.E. (1963) The distribution of colony-forming cells among spleen colonies. *J. Cell Physiol.*, **62**, 327–336.

5 Martin, G.R. (1981) Isolation of a pluripotent cell line from early mouse embryos cultured in medium conditioned by teratocarcinoma stem cells. *Proc. Natl. Acad. Sci. USA*, **78** (12), 7634–7638.

6 Evans, M.J. and Kaufman, M.H. (1981) Establishment in culture of pluripotential cells from mouse embryos. *Nature*, **292**, 154–156.

7 Thomson, J.A., Itskovitz-Eldor, J., Shapiro, S.S., Waknitz, M.A., Swiergiel, J.J., Marshall, V.S., and Jones, J.M. (1998) Embryonic stem cell lines derived from human blastocysts. *Science*, **282**, 1145–1147.

8 Gärtner, W., Gröbler, B., Schubert, D., Wabnitz, H., and Wilhelmi, B. (1988) Fluorescence scanning microscopy combined with subnanosecond time resolution. *Exp. Tech. Phys.*, **36**, 443–451.

9 König, K. (1989) PhD thesis. Optical cancer diagnosis and picosecond fluorescence lifetime microscopy. Archive of the Friedrich Schiller University Jena.

10 Bugiel, I., König, K., and Wabnitz, H. (1989) Investigations of cells by fluorescence laser scanning microscopy with subnanosecond time resolution. *Laser Life Sci.*, **3**, 47–53.

11 Hänninen, P.E., Soini, E., and Hell, S.W. (1994) Continuous wave excitation two-photon fluorescence microscopy. *J. Microsc.*, **176**, 222–225.

12 König, K., Liang, H., Berns, M.W., and Tromberg, B.J. (1995) Cell damage by near-IR microbeams. *Nature*, **377**, 20–21.

13 König, K., Liang, H., Berns, M.W., and Tromberg, B.J. (1996) Cell damage in near infrared multimode optical traps as a result of multiphoton absorption. *Opt. Lett.*, **21**, 1090–1092.

14 König, K. (2000) Multiphoton microscopy in life sciences. *J. Microsc.*, **200**, 83–104.

15 Squirrel, J.M., Wokosin, D.L., White, J.G., and Bavister, B.D. (1999) Long-term two-photon fluorescence imaging of mammalian embryos without compromising viability. *Nat. Biotechnol.*, **17**, 763–767.

16 Fischer, F., Beate, V., Puschmann, S., Greinert, R., Breitbart, E., Kiefer, J., and Wepf, R. (2008) Assessing the risk of skin damage due to femtosecond laser irradiation. *J. Biophoton.*, **1** (6), 470–477.

17 Rebecca, M.W., Warren, R.Z., and Watt, W.W. (2001) Multiphoton microscopy in biological research. *Curr. Opin. Chem. Biol.*, **5**, 603–608.

18 Zipfel, W.R., Williams, R.M., and Webb, W.W. (2003) Nonlinear magic: multiphoton microscopy in the biosciences. *Nat. Biotechnol.*, **21**, 1369–1377.

19 König, K. and Schneckenburger, H. (1994) Laser-induced autofluorescence for medical diagnosis. *J. Fluoresc.*, **4**, 17–40.

20 Freund, I. and Deutsch, M. (1996) 2nd harmonic microscopy of biological tissue. *Opt. Lett.*, **11**, 94–96.

21 Masters, B.R., So, P.T., and Gratton, E. (1997) Multiphoton excitation fluorescence microscopy and spectroscopy of *in vivo* human skin. *Biophys. J.*, **72**, 2405–2412.

22 Zoumi, A., Yen, A., and Tromberg, B.J. (2002) Imaging cells and extracellular matrix in vivo by using second harmonic generation and two-photon excited fluorescence. *Proc. Natl. Acad. Sci. USA*, **99**, 11014–11019.

23 König, K., Schenke-Layland, K., Riemann, I., and Stock, U.A. (2005) Multiphoton autofluorescence imaging of

intratissue elastic fibers. *Biomaterials*, **26**, 495–500.

24 Cox, G., Moreno, N., and Feijo, J. (2005) Second harmonic imaging of plant polysaccharides. *J. Biomed. Opt.*, **10**, 024013.

25 Chance, B., Schoener, B., Oshino, R., Itshak, F., and Nakase, Y. (1979) Oxidation-reduction ratio studies of mitochondria in freeze-trapped samples. *J. Biol. Chem.*, **254**, 4764–4771.

26 Huang, S., Ahmed, A., Heikal, A., and Webb, W.W. (2002) Two-photon fluorescence spectroscopy and microscopy of NAD(P)H and flavoproteins. *Biophys. J.*, **82**, 2811–2825.

27 Niesner, R., Peker, B., Schlüsche, P., and Gericke, K.H. (2004) Noniterative bi-exponential fluorescence lifetime imaging in the investigation of cellular metabolism by mean of NAD(P)H autofluorescence. *Phys. Chem. Chem. Phys.*, **5** (8), 1141–1149.

28 Dimitrow, E., Riemann, I., Ehles, A., Koehler, J., Norgauer, J., Elsner, P., König, K., and Kaatz, M. (2009) Spectral fluorescence lifetime detection and selective melanin imaging by multiphoton laser tomography for melanoma diagnosis. *Exp. Dermatol.*, **18** (6), 509–515.

29 Vogel, A., Noack, J., Hüttman, G., and Paltauf, G. (2005) Mechanisms of femtosecond laser nanosurgery of cells and tissues. *Appl. Phys. B*, **81**, 1015–1047.

30 König, K., Becker, T.W., Fischer, P., Riemann, I., and Halbhuber, K.J. (1999) Pulse-length dependence of cellular response to intense near-infrared laser pulses in multiphoton microscopes. *Opt. Lett.*, **24**, 113–115.

31 König, K., Riemann, I., and Fritzsche, W. (2001) Nanodissection of human chromosomes with near infrared femtosecond laser pulses. *Opt. Lett.*, **26**, 819–821.

32 König, K., Riemann, I., Fischer, P., and Halbhuber, K.J. (1999) Intracellular nanosurgery with near infrared femtosecond laser pulses. *Cell Mol. Biol.*, **45**, 195–201.

33 König, K. (2000) Robert Feulgen prize lecture 2000. Laser tweezers and multiphoton microscopes in life sciences. *Histochem. Cell Biol.*, **114**, 79–92.

34 König, K., Riemann, I., Stracke, F., and Le Harzic, R. (2005) Nanoprocessing with nanojoule near-infrared femtosecond laser pulses. *Med. Laser Appl.*, **20**, 169–184.

35 Tirlapur, U.K. and König, K. (2002) Targeted transfection by femtosecond laser. *Nature*, **418**, 4295–4298.

36 Watanabe, W., Matsunaga, S., Shimada, T., Higashi, T., Fukui, K., and Itoh, K. (2005) Femtosecond laser disruption of mitochondria in living cells. *Med. Laser Appl.*, **20**, 185–191.

37 Heisterkamp, A., Maxwell, I.Z., Mazur, E., Underwood, J.M., Nickerson, J.A., Kumar, S., and Ingber, D.E. (2005) Pulse energy dependence of subcellular dissection by femtosecond laser pulses. *Opt. Express*, **13**, 3690–3696.

38 Shimada, T., Watanabe, W., Matsunaga, S., Higashi, T., Ishii, H., Fukui, K., Isobe, K., and Itoh, K. (2005) Intracellular disruption of mitochondria in a living HeLa cell with a 76-MHz femtosecond laser oscillator. *Opt. Express*, **13**, 9869–9880.

39 Sacconi, L., Tolic-Norrelykke, I., Antolini, R., and Pavone, F.S. (2005) Combined intracellular three-dimensional imaging and selective nanosurgery by a nonlinear microscope. *J. Biomed. Opt.*, **10**, 14002.

40 König, K., Krauss, O., and Riemann, I. (2002) Intratissue surgery with 80MHz nanojoule femtosecond laser pulses in the near infrared. *Opt. Express*, **10**, 171–176.

41 Pittenger, M.F., Mackay, A.M., Beck, S.C., Jaiswal, R.K., Douglas, R., Moska, J.D., Moorman, M.A., Simonetti, D.W., Craig, S., and Marshak, D.R. (1999) Multilineage potential of adult human mesenchymal stem cells. *Science*, **284**, 143–147.

42 Sakaguchi, Y., Sekiya, I., Yagishita, K., and Muneta, T. (2005) Comparison of human stem cells derived from various mesenchymal tissues: superiority of synovium as a cell source. *Arthritis Rheum.*, **52**, 2521–2529.

43 Sauer, H., Rahimi, G., Hescheler, J., and Wartenberg, M. (1999) Effects of electrical fields on cardiomyocyte differentiation of embryonic stem cells. *J. Cell Biochem.*, **75**, 710–723.

44 Zou, G.-M., Chen, J.-J., and Ni, J. (2006) Light induces differentiation of mouse embryonic stem cells associated with

activation of ERK5. *Oncogene*, **25**, 463–469.

45 Wagner, D.R., Lindsey, D.P., Li, K.W., Tummala, P., Chandran, S.E., Smith, R.L., Longaker, M.T., Carter, D.R., and Beaupre, G.S. (2008) Hydrostatic pressure enhances chondrogenic differentiation of human bone marrow stromal cells in osteochondrogenic medium. *Ann. Biomed. Eng.*, **36**, 813–820.

46 Ge, D., Liu, X., Wu, J., Tu, Q., Shi, Y., and Chen, H. (2009) Chemical and physical stimuli induce cardiomyocyte differentiation from stem cells. *Biochem. Biophys. Res. Commun.*, **381**, 317–321.

47 Suwa, T., Yang, L., and Hornsby, P.J. (2001) Telomerase activity in primary cultures of normal adrenocortical cells. *J. Endocrinol.*, **170**, 677–684.

48 Oh, B.-K., Lee, C.-H., Park, C., and Park, Y.N. (2004) Telomerase regulation and progressive telomere shortening of rat hepatic stem-like epithelial cells during *in vitro* aging. *Exp. Cell Res.*, **298**, 445–454.

49 Ju, Z. and Rudolph, K.L. (2006) Telomeres and telomerase in stem cells during aging and disease. *Genome Dyn.*, **1**, 84–103.

50 Rubio, D., Garcia-Castro, J., Martin, M.C., Fuente, R., Cigudosa, J., Lloyd, A.C., and Bernad, A. (2005) Spontaneous human adult stem cell transformation. *Cancer Res.*, **65**, 3035–3039.

51 Bruder, S.P., Jaiswal, N., and Haynesworth, S.E. (1997) Growth kinetics, selfrenewal, and the osteogenic potential of purified human mesenchymal stem cells during extensive subcultivation and following cryopreservation. *J. Cell Biochem.*, **64**, 278–294.

52 Chung, S., Shin, B.-S., Hedlund, E., Pruszak, J., Ferree, A., Kang, U.J., Isacson, O., and Kim, K.-S. (2006) Genetic selection of sox1GFP-expressing neural precursors removes residual tumorigenic pluripotent stem cells and attenuates tumor formation after transplantation. *J. Neurochem.*, **97**, 1467–1480.

53 Schenke-Layland, K., Riemann, I., Damour, O., Stock, U.A., and König, K. (2006) Two-photon microscopes and *in vivo* multiphoton tomographs - Powerful diagnostic tools for tissue engineering and drug delivery. *Adv. Drug Deliv. Rev.*, **58**, 878–896.

54 Takahashi, K., Tanabe, K., Ohnuki, M., Narita, M., Ichisaka, T., Tomoda, K., and Yamanaka, S. (2007) Induction of pluripotent stem cells from adult human fibroblasts by defined factors. *Cell*, **131**, 861–872.

55 Becker, W. (ed.) (2005) *Advanced Time-Correlated Single-Photon Counting Techniques*, Springer, Berlin, Heidelberg, New York.

56 Periasamy, A. (ed.) (2001) *Methods in Cellular Imaging*, Oxford University Press, Oxford, New York.

57 Periasamy, A. and Clegg, R.M. (eds) (2009) *FLIM Microscopy in Biology and Medicine*, Taylor and Francis Group (CRC Press), Roca Raton, London, New York.

58 Diskinson, M.E., Bearman, G., Tille, S., Landsford, R., and Fraser, S.E. (2001) Multi-spectral imaging linear unmixing add a whole new dimension to laser scanning fluorescence microscopy. *BioTechniques*, **31**, 1272–1278.

59 Uchugonova, A. and König, K. (2008) Two-photon autofluorescence and second-harmonic imaging of adult stem cells. *J. Biomed. Opt.*, **13**, 054068.

60 Uchugonova, A., Isemann, A., Gorjup, E., Tempea, G., Bückle, R., Watanabe, W., and König, K. (2008) Optical knock out of stem cells with extremely ultrashort femtosecond laser pulses. *J. Biophoton.*, **1**, 463–469.

61 Guo, H.-W., Chen, C.-T., Wei, Y.-H., Lee, O.K., Gukassyan, V., Kao, F.-J., and Wang, H.-W. (2008) Reduced nicotinamide adenine dinucleotide fluorescence lifetime separates human mesenchymal stem cells from differentiated progenies. *J. Biomed. Optic. Lett.*, **13**, 050505(1-3).

62 Lee, H.S., Teng, S.W., Chen, H.C., Lo, W., Sun, Y., Lin, T.Y., Chiou, L.L., Jiang, C.C., and Dong, C.Y. (2006) Imaging the bone marrow stem cells morphogenesis in PGA scaffold by multiphoton autofluorescence and second harmonic (SHG) imaging. *Tissue Eng.*, **12**, 2835–2842.

63 Teisanu, R.M., Lagasse, E., Whitesides, J.F., and Stripp, B.R. (2009) Prospective isolation of bronchial stem cells based upon immunophenotypic and autofluorescence characteristic. *Stem Cells*, **27**, 612–622.

64 Wu, B.P., Tao, Q., and Lyle, S. (2005) Autofluorescence in the stem cell region of the hair follicle bulge. *J. Invest. Dermatol.,* **124**, 860–862.

65 Zhao, L.-R. and Nam, S.C. (2007) Multiphoton microscope imaging: the behavior of neural progenitor cells in the rostral migratory stream. *Neurosci. Lett.,* **425**, 83–88.

66 König, K., Weinigel, M., Hoppert, D., Bückle, R., Schubert, H., Köhler, M.J., Kaatz, M., and Elsner, P. (2008) Multiphoton tissue imaging using high-NA microendoscopes and flexible scan heads for clinical studies and small animal research. *J. Biophoton.,* **1**, 506–513.

67 Uchugonova, A., König, K., Bückle, R., Isemann, A., and Tempea, G. (2008) Targeted transfection of stem cells with sub-20fs laser pulses. *Opt. Express,* **16**, 9357–9364.

68 Uchugonova, A., Müller, J., Bückle, R., Tempea, G., Isemann, A., Stingl, A., and König, K. (2008) Negatively-chirped laser enables nonlinear excitation and nanoprocessing with sub 20fs pulses. *Proc. of SPIE,* **6860**, 686015.

69 Földes-Papp, Z., König, K., Studier, H., Bückle, R., Breunig, H.G., Uchugonova, A., and Kostner, G.M. (2009) Trafficking of mature miRNA into the nucleus of life liver cells. *Curr. Pharm. Biotechnol.,* **10** (6) 569–578.

70 Tsampoula, X., Garces-Chavez, V., Comrie, M., Stevenson, D.J., Agate, B., Brown, C.T.A., Gunn-Moore, F., and Dholakia, K. (2007) Femtosecond cellular transfection using a nondiffracting light beam. *Appl. Phys. Lett.,* **91**, 053902.

2
In Vivo Nanosurgery

Leonardo Sacconi and Francesco S. Pavone

2.1
Introduction

The linkage between optics and biology has always been extremely strong. In fact, it was through the use of the first optical microscopes in the seventeenth century that fundamental concepts such as the existence of the cell itself were defined. Anton van Leeuwenhoek first saw and described bacteria and the circulation of blood corpuscles in capillaries; Robert Hooke coined the term "cell" in 1665 after observing with his microscope a sliver of cork.

In the 1960s, the advent of the laser totally revolutionized biology and medicine. Currently, we can monitor a virus infecting a cell, cauterize an ulcer in a patient without making a single incision, or image blood flow inside a mouse brain during a stroke. Recently, femtosecond lasers have gained ground due to their versatile application to a large variety of problems. The first application developed for short-pulsed lasers in biology was multiphoton fluorescence microscopy [1, 2]. The nonlinear nature of multiphoton processes provides an absorption volume spatially confined to the focal region. The localization of the excitation is maintained even in strongly scattering tissues, allowing deep, high-resolution *in vivo* microscopy [3, 4].

In parallel to its application in imaging, multiphoton absorption has also been used as a tool for the selective disruption of cellular and intracellular labeled structures [5–7]. A similar approach has also been applied *in vivo*, where two-photon imaging and laser-induced lesions have been combined. Femtosecond laser ablations have been applied with great success, for example, in studying the complex events associated with embryo development *in vivo* [8]. Laser-induced lesions have also been used to breach the blood–brain barrier and demonstrate the surveillant role of microglia [9] and produce targeted photodisruptions as a model of stroke [10]. As a means to automate the three-dimensional histological analysis of brain tissue, Tsai *et al.* [11] have demonstrated the use of femtosecond laser pulses to iteratively cut and image both fixed and fresh tissue. Other groups have taken advantage of multiphoton absorption to ablate or dissect individual neurons. Multiphoton nanosurgery has been performed in worms to study axon regeneration [12] and

Laser Imaging and Manipulation in Cell Biology. Edited by Francesco S. Pavone
Copyright © 2010 WILEY-VCH Verlag GmbH & Co. KGaA, Weinheim
ISBN: 978-3-527-40929-7

dissect the role of specific neurons in behavior [13]. More recently, Sacconi *et al.* [14] demonstrated a method for performing multiphoton nanosurgery in the central nervous system of mice.

In this chapter we present some basic features of multiphoton nanosurgery and show some examples to illustrate the advantages offered by this novel methodology. We briefly introduce the physical mechanisms of femtosecond laser ablation (Section 2.2) and describe a typical experimental system used to perform multiphoton nanosurgery (Section 2.3). We then review some applications of femtosecond laser to the disruption of intracellular structures in living cells and to the induction of lesions *in vivo*.

2.2
Physical Mechanisms

The underlying physical mechanisms of femtosecond laser nanosurgery on cells and tissues are not easily understood. In general, we must consider three potential mechanisms for the production of damage in the target structure: (i) photochemical processes due to multiphoton absorption; (ii) generation of large thermoelastic stresses; and (iii) thermal, mechanical, and chemical processes emanating from optical breakdown (plasma formation) produced by a combination of multiphoton and cascade ionization processes [15, 16]. The irradiance thresholds of these damage mechanisms have been estimated by Vogel *et al.* [16] and are summarized in Table 2.1. As reported, above a certain laser irradiance (0.26×10^{12} W cm^{-2}) the absorption of femtosecond pulses produces free electrons. The free electrons are generated at a relatively large range of irradiances below the optical breakdown threshold and with a deterministic relationship between free-electron density and irradiance. This provides a large "tuning range" for the creation of spatially confined chemical, thermal, and mechanical effects via free electron generation. For simplicity, two parameter regimes can be established for nanosurgery. One technique uses a long series of pulses from fs oscillators with repetition rates of the order of 80 MHz and pulse energies well below the optical breakdown threshold. In this case, the energies employed rarely exceed those used in nonlinear imaging [5, 6, 14, 17–21]. The other

Table 2.1 Three potential mechanisms for the production of damage in the target structure; the irradiation threshold and the free electron density were estimated for 100 fs, 800 nm laser pulses focused with a 1.3 NA objective [16].

Damage mechanism	Irradiance threshold (W cm^{-2})	Electron density per pulse (cm^{-3})
Photochemical process	0.26×10^{12}	2.1×10^{13}
Thermoelastic stresses	5.1×10^{12}	0.24×10^{21}
Optical breakdown	6.54×10^{12}	1.0×10^{21}

approach uses amplified series of pulses at repetition rates of 1 kHz and pulse energies slightly above the threshold [7, 12, 22–25].

The mechanisms on which these two regimes rely are quite different. The low repetition rate regime effects are related to thermoelastically-induced formation of transient cavities. An extensive review focused on this regime has been recently published by Tsai *et al.* [26]. On the other hand, in the low-density plasma regime dissection is mediated by free-electron-induced chemical decomposition. The chemical decomposition can be induced through two different mechanisms: 1) the fragmentation of the biomolecule derives from its resonant interaction with quasi-free electrons: the capture of a quasi-free electron in an antibonding molecular orbital causes the rupture of a chemical bond and determines the degradation of the molecule; 2) the fragmentation of the biomolecule derives from its interaction with reactive oxygen species (ROS, i.e. free radicals that contain the oxygen atom), originated from laser-induced ionization and dissociation of water molecules.

2.3
Experimental Setup

In this section we describe a typical experimental setup that combines multiphoton imaging and nanosurgery. This particular experimental system was used in Pavone's group to perform *in vivo* multiphoton nanosurgery on cortical neurons [14]. The setup consists of a custom-made upright two-photon microscope. A mode-locked Ti: sapphire Chameleon laser (Coherent Inc) provided the excitation light, which consists of 120 fs width pulses at a 90 MHz repetition rate. The scanning head is made up of two closed-loop feedback galvanometer mirrors VM500 (GSI Lumonics), rotated about orthogonal pivots and coupled by a pair of spherical mirrors. A scanning lens and a microscope tube lens expanded the beam before it was focused onto the specimen by the objective lens XLUM 20×, NA 0.95, WD 2 mm (Olympus). A closed loop piezoelectric stage P-721 (Physik Instrumente) allowed axial displacements of the objective up to 100 µm with nanometric precision. Fluorescence light was separated from the laser optical path by a dichroic beam splitter 685DCXRU (Chroma Technology) positioned as close as possible to the objective lens (non-descanning mode) and was focused onto a photomultiplier detector H7710-13 (Hamamatsu Photonics). A two-photon florescence cut-off filter, E700SP-2P (Chroma Technology), eliminated reflected laser light. All the microscope optics were fixed onto a custom vertical honeycomb steel breadboard. The electronic components of the setup were computer-controlled during microscopy and nanosurgery experiments. Custom-made software was developed in LabVIEW 7.1 (National Instruments). The excitation wavelength was chosen to maximize the multiphoton absorption of the labeled structure.

The system was designed for maximum stability during laser neurosurgery by incorporating a closed-loop feedback system to precisely control beam positioning in the *x*, *y* and *z* axes. First, a *z*-stack was acquired with two-photon microscopy to obtain a 3D-reconstruction of the labeled sample and select an *x*-, *y*-, and *z*-coordinate for our desired lesion site. Second, laser-dissection was carried out by parking the laser beam at the chosen point and increasing the power ($\approx 5\times$ more than the laser power used for imaging; [6, 14]). The dose of laser energy was set by opening the shutter for a prescribed period with an exposure ranging from 150 to 300 ms [6, 14]. Finally, the laser power was decreased back and a 3D image of the sample was acquired to visualize the effect of laser irradiation.

2.4
Subcellular Nanosurgery

The first published report of femtosecond laser subcellular nanosurgery was by Koenig *et al.* in 1999 [17], who demonstrated the potential of this technique by ablating nanometer-sized regions of the genome within the nucleus of living cells (Figure 2.1a). Such intranuclear nanosurgery was made possible by the application of highly intense near-infrared femtosecond laser pulses. The destructive multiphoton effect was based on $10^{12}\,\mathrm{W\,cm^{-2}}$ light intensities and limited to a sub-femtoliter focal volume of a high numerical aperture objective. A minimum cut size of ~100 nm (which is below the diffraction-limited spot size) was achieved during this partial dissection of a chromosome, corresponding to a minimum material removal of ~0.003 μm^3, as determined by scanning-force microscopy [5]. Furthermore, as shown in Figure 2.1b, complete dissection could be performed with FWHM cut sizes below 200 nm. Inspired by these seminal works, Sacconi *et al.* used near-IR femtosecond laser pulses for a combination of microscopy and nanosurgery on fluorescently labeled structures within living cells [6]. In this work microtubule structures tagged with green fluorescent protein (GFP) were dissected employing the same laser beam used for two-photon imaging. Figure 2.2a shows an example of nanosurgery on the mitotic spindle of the yeast *Schizosaccharomyces pombe* in anaphase B. The spindle shown in the figure was irradiated in the middle just after the acquisition of the first image, after which it bent and subsequently broke into two segments that crossed each other, forming an X-shaped structure. The nanosurgery was also performed on the same type of cell, but with the nuclear membrane labeled with GFP. Since the nuclear membrane encloses the spindle tightly, it is possible to know the position of the spindle in the cells with a labeled nuclear membrane, without labeling the spindle directly. Irradiation performed on the unlabeled spindle in these cells did not induce any variation in the shape of the nuclear membrane enclosing the spindle, which implies that the spindle shape did not change either (Figure 2.2b). This result suggested that the laser nanosurgery was due to multi-photon absorption by the fluorescence marker. This important aspect can be exploited to increase the specificity and spatial resolution of this methodology. Tolic-Norrelykke *et al.* [27] have used this methodology to investigate the nature of the force that astral microtubules exert on the spindle. They performed two sets of laser nanosurgery experiments. In the first experiment, spindles were severed in their midzone. The rationale behind this experiment was that a spindle that is

(a)

(b)

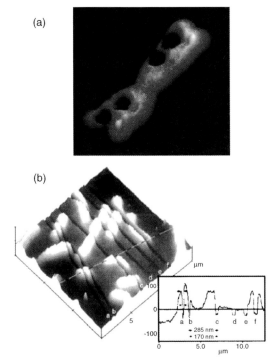

Figure 2.1 Femtosecond laser nanodis section of chromosomes. (a) Holes are drilled in the chromosome by fixing the laser on a spot. (b) Complete dissection of human chromosomes with 800-nm sub-nJ femtosecond laser pulses. The depth profile of cut a indicates a FWHM cut size below 200 nm. The other cuts (b–f) are in the range 200–400 nm, which is still less than 40% of the diffraction-limited spot diameter. The zero line corresponds closely to the lower surface of the chromosomes, whereas the negative values reveal cuts in the mprotein matrix. The numbers of line scans are a, b, 500; c, 1000; d, 2000; e, f, 2500. Adapted from References [5,17].

weakened in its midzone, or broken into two separate parts, should provide less resistance to astral forces than an intact spindle. They observed that in either weakened or broken spindles with a single astral microtubule extending to the cortex, the adjacent spindle arm rotated away from the contact site between the tip of the astral microtubule and the cell cortex. The second experiment was performed on cells in which two astral microtubules extended simultaneously from a single spindle pole body in opposite directions. They dissected one of the astral microtubules and observed the behavior of the spindle. Again they found that the spindle rotated away from the contact site between the tip of the astral microtubule and the cell cortex. Taken together, these data suggest that astral microtubules exert pushing forces on the mitotic spindle in *S. pombe*, thereby aligning the spindle with the long axis of the cell.

In the previous examples, laser dissections were achieved using a long series of pulses from fs oscillators with repetition rates of the order of 80MHz and pulse energies well below the optical breakdown threshold (Section 2.2). The first

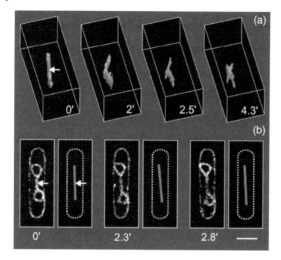

Figure 2.2 Nanosurgery on the mitotic spindle. (a) The spindle was irradiated in the middle just after acquisition of the first image, after which it bent and subsequently broke into two segments that crossed each other to form an X-shaped structure. (b) Three images are shown, each representing a single optical section. Time in minutes is noted. To the right of each of the images is a scheme showing the position of the spindle (solid line) and the cell membrane (dotted line). Nanosurgery was performed on the center of the spindle after acquisition of the first image. The nanosurgery did not induce any variation in the shape of the nuclear membrane enclosing the spindle, which suggests that the spindle shape did not change either. Scale bar is 4 μm. Adapted from Reference [6].

subcellular organelle disruption with low repetition rate pulses was presented by Shen *et al.* in 2005 [23]. In this work, a single labeled mitochondrion was disrupted in a living cell. Figure 2.3 shows the ablation of a single mitochondrion, about 5 μm in length and separated by less than 1 μm from multiple neighboring mitochondria. After irradiating a fixed spot on the organelle with a few hundred 2 nJ laser pulses at a

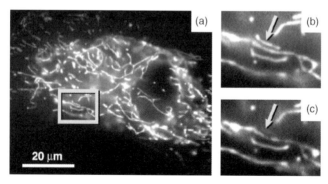

Figure 2.3 Ablation of a single mitochondrion in a living endothelial cell. (a) Fluorescence microscope image showing multiple mitochondria before femtosecond laser irradiation. Target mitochondrion (marked by arrow) (b) before and (c) after laser ablation with 2-nJ pulses. Adapted from Reference [23].

1 kHz repetition rate, the entire mitochondrion disappeared from the image, whereas neighboring mitochondria were not affected by the irradiation. This result provided direct evidence that mitochondria do not form an interconnected network.

Other groups have addressed questions regarding mitochondrial ablation in living cells. Watanabe and Arakawa [7] have ablated mitochondria using high repetition laser pulses and studied their fragmentation. They also observed that cells maintain full viability, undergoing normal division following such laser manipulations. Furthermore, Shimada *et al.* [28] have reported the intracellular disruption of mitochondria in a living cell with a 76-MHz femtosecond laser oscillator. By comparison of these studies, clearly, the specificity of the laser ablation obtained is comparable when using either low or high pulse repetition rates.

Finally, Kumar *et al.* [29] have used laser nanoscissors to confirm that stress fibers behave as viscoelastic cables that are tensed through the action of actomyosin motors. In addition, they quantified the retraction kinetics of stress fibers *in situ*, and explored their contribution to the overall mechanical stability of the cell and interconnected extracellular matrix.

2.5
In Vivo Nanosurgery

The spatial localization of multiphoton excitation and the high penetration depth of the IR laser light have been exploited to perform selective lesions *in vivo*. The nematode worm *Caenorhabditis elegans* is particularly well suited to femtosecond laser nanosurgery due to its transparency and size. Moreover, the neuronal network of *C. elegans* totals only 302 neurons and their connectivity has been mapped [30]. Study of the individual neuronal function had previously been limited to observing behavioral changes in mutant worms. Femtosecond laser dissection enables us to directly observe the behavioral function of individual neurons by severing neuronal fibers, similar to cutting wires in an electrical circuit. The first published report of femtosecond neuron ablation was by Yanik *et al.* in 2004 [12]. In this work the authors performed a femtosecond laser ablation of the axon that controls the crawling motion of the animal (Figure 2.4a). After the laser nanosurgery the backward crawl of the nematode was greatly hindered. Remarkably, however, the locomotion of the worm gradually returned normal within 24 h in all the worms (Figure 2.4c). Some of the axons also exhibited partial or full regrowth when they were imaged again (Figure 2.4b).

The function of a neuron or neuronal part can be identified by observing behavioral changes after the surgery. An example of this is the study by Chung *et al.* [13] on the thermotactic behavior of the nematodes. The *C. elegans* retain a memory about the temperature at which they were cultivated, and when placed at a higher temperature they crawl down the gradient toward the cultivation temperature. Through genetic mutations, the AFD neuron was implicated in this cryophilic behavior, but the exact sensing mechanism and signal processing remained unknown. Through femtosecond laser nanosurgery it is possible to ablate the neuronal cell bodies and the sensory

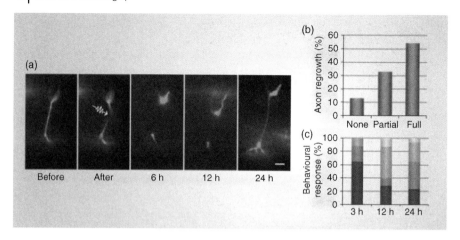

Figure 2.4 Femtosecond laser axotomy in C. elegans worms using 100 pulses of low energy (40 nJ) and short duration (200 fs) and a repetition rate of 1 kHz. (a) Fluorescence images of axons labeled with green fluorescent protein before, immediately after, and in the hours following axotomy. The arrow indicates point of severance. Scale bar, 5 μm. (b) Statistics of axon growth 24 h after axotomy, based on fluorescence images ($n = 52$ axons). (c) Time-course analysis of backward motion of worms following axotomy. Seventeen worms were scored blindly at different time points. Improvement in backward motion was graded as four levels from shrinker behavior (bottom segment) up to wild-type behavior (top segment) in the hours following axotomy. Adapted from Reference [12].

dendrites to investigate their comparative contribution to the cryophilic movement. Severing the dendrites in young adult worms permanently abolishes these dendrites' sensory contribution. The AFD neuron also regulates the cryophilic bias of the worm, but there is no evidence that it does so by generating an opposing thermophilic behavior. By disrupting other neurons, interconnected with the AFD it was possible to isolate the source of the cryophilic behavior to the AFD neuron.

As reported in these examples, multiphoton nanosurgery has been performed in worms to study axon regeneration and dissect the role of specific neurons in behavior. Nevertheless, the potential of this technique has not yet been fully explored in the mammalian central nervous system. Consequently, the investigation of many outstanding problems in neurobiology and human neurodegenerative disease still appears inaccessible and may benefit from this approach. More recently, Sacconi et al. [14] have demonstrated a method for performing multiphoton nanosurgery in the central nervous system of mice. In this work they exploited the spatial localization of multiphoton excitation to perform selective lesions on the neuronal processes of cortical neurons in living mice expressing fluorescent proteins. Neurons were irradiated with a focused, controlled dose of femtosecond laser energy delivered through cranial optical windows. The morphological consequences were then characterized with time lapse 3D two-photon imaging over a period of minutes to days after the procedure. This methodology was applied to dissect single dendrites with sub-micrometric precision without causing any visible collateral damage to the surrounding neuronal structures (Figure 2.5a). The spatial precision of this method

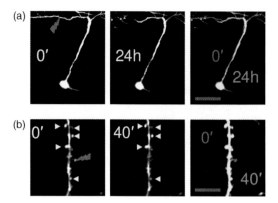

Figure 2.5 *In vivo* multiphoton nanosurgery on cortical neurons. (a) Maximum-intensity z-projection (from 500 to 100 μm depth) of a layer-5 pyramidal neuron before and 24 h after dendritic dissection. The dendrite was irradiated as indicated by the tip of the lightning. The last panel shows an overlay of the neuron before and after laser dissection. This merge shows the integrity of the remaining structure after dissection. Scale bar, 60 μm. (b) Time-lapse images of a section of the dendrite where several spines are present. The single spine indicated by the lightning symbol was irradiated just after the acquisition of the first image. The arrowheads highlight the stability of the surrounding dendritic spines. The last panel shows an overlay of the dendrite before and after spine dissection. This panel emphasizes the spine stability and the absence of any swelling in the dendrite. Scale bar, 15 μm. Adapted from Reference [14].

was demonstrated by ablating individual dendritic spines, while sparing the adjacent spines and the structural integrity of the dendrite (Figure 2.5b).

Another application of femtosecond laser pulses is imaging blood flow through cortical blood vessels and implementing a model for stroke by disrupting subsurface blood vessels in living animals [9]. Figure 2.6a and b show an optical three-dimensional reconstruction of vasculature in the cortex of a living mouse. The vasculature was imaged and the blood flow rate was quantified through two-photon microscopy. High-fluence, femtosecond laser pulses were focused on a blood vessel and the laser parameters were varied to achieve three different forms of vascular insult (Figure 2.6c). First, vessel rupture was induced at the highest optical energies, which provided a model of hemorrhage. Second, extravasation (blood vessel leakage) of blood components was possible by focusing low energy pulses, while still maintaining blood flow through the vessel. Finally, a blood clot was formed by further irradiation of the extravasated blood vessel. The blood flow through the downstream vessels was imaged in real time before and after the induced infraction. The speed of flow was shown to drop significantly after the stroke was induced. This demonstrated that blockage of a single microvessel can lead to local cortical ischemia.

Femtosecond laser ablation can be applied to larger scale systems such as whole tissues. One application that utilizes the versatility of femtosecond lasers is an all-optical histology. Tsai *et al.* [11] showed that by combining two-photon microscopy

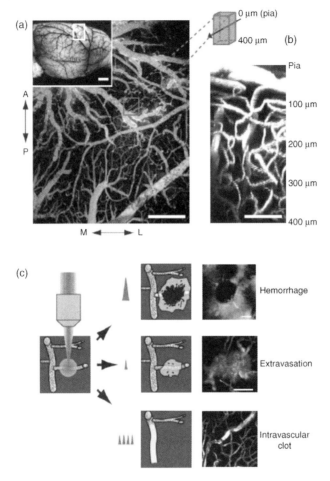

Figure 2.6 Targeted insult to subsurface cortical blood vessels using ultrashort laser pulses: three models of stroke. (a,b) Maps of fluorescein-dextran-labeled vasculature of rat parietal cortex. Inset in (a) shows latex-filled surface arteries and arterioles in rat cortex, and the white rectangle indicates the approximate location of the craniotomy. The images in (a) are maximal projections along the optical axis of near-surface vasculature. Scale bars: (a) 500 μm (inset 5 mm), (b) 100 μm. (c) Schematic of the three different vascular lesions that are produced by varying the energy and number of laser pulses. At high energies, photodisruption produces hemorrhages, in which the target is ruptured, blood invades the brain tissue, and a mass of RBCs form a hemorrhagic core. At low energies the target vessel remains intact, but transiently leaks blood plasma and RVCs forming extravasation. Multiple pulses at low energy lead to thrombosis that can completely occlude the target vessel, forming an intravascular clot. Scale bar, 50 μm. Adapted from Reference [10].

and surface laser ablation they can create a three-dimensional map of mouse cortical tissue. This is a two-step process: in the first step the specimen is labeled and a thin slice is imaged through multiphoton microscopy; in the second step, the imaged section is ablated and removed, and the process is repeated. As an example they performed serial ablation and imaging of the fixed neocortex of mice in which infragranular projection neurons selectively expressed YFP. Optical imaging and ablation was performed in the radial direction over a lateral extent of 200 µm. Each iteration of imaging consisted of a total depth of approximately 110 µm, of which

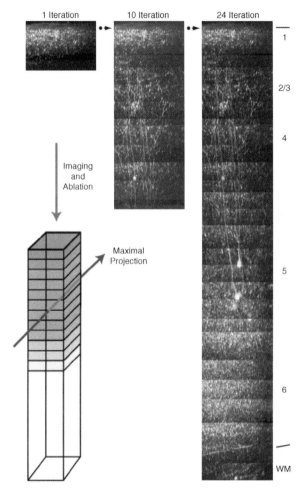

Figure 2.7 All-optical histology using ultrashort laser pulses. Iterative processing of a block of neocortex of a YFP labeled transgenic mouse. Twenty-four successive cutting and imaging cycles are shown. The laser was focused onto the cut face with a 20× magnification, 0.5 NA water objective, and single passes, at a scan rate of 4 mm s^{-1}, were made to optically ablate successive planes at a depth of 10 µm each with total thicknesses between 40 and 70 µm per cut. The energy per pulse was maintained at 8 µJ. Each stack of images represents a maximal side projection of all accumulated optical sections obtained using a two-photon laser scanning microscope at an excitation wavelength of 920 nm. The sharp breaks in the images shown in successive panels demarcate the cut boundaries. Adapted from Reference [11].

60 µm represented new information and 50 µm represented overlap with prior images as a means to cross-check alignment. The image stacks are displayed as a maximal projection in the coronal direction (Figure 2.7). The stacks from 24 iterations of cutting and imaging were overlaid and merged to generate a 3D matrix of intensity values that extend the full depth of neocortex. The maximal projection of this matrix allows visualization of fine structures and compares favorably with the published coronal images [31]. This technique allows for large sections of tissues to be reconstructed. It has an important advantage over typical pathological methods in using fresh, rather than frozen, tissue. The laser ablation also has higher precision than mechanical knives and induces smaller additional damage to the tissue during slicing.

2.6
Conclusions

Laser nanosurgery is an important tool in biology for the dissection or deletion of specific structures without using genetic methods or chemical agents [32]. Different laser systems, such as picosecond, nanosecond, and UV lasers, have been used for subcellular disruption [33–39]. Owing to the shorter wavelengths, the disruption of these lasers systems are limited to surfaces or thin samples, including a single cell in culture. Recently, multiphoton absorption has reinvigorated this area of research and become a useful tool for the selective disruption and dissection of cellular structures in living cells and *in vivo*. By varying the laser repetition rate, pulse energy, number of pulses irradiated and the focusing conditions, femtosecond nanosurgery can be performed with different interaction regimes and work both on the subcellular and the tissue level with unprecedented specificity. As illustrated by the few examples described in this chapter, the impact of multiphoton nanosurgery on modern biology (especially, but not only, cell biology and neurobiology) has already been remarkable. The capability of combining visualization with selective and direct disruption of intracellular structures offers an extraordinary tool for the investigation of many vital processes, including cellular division, locomotion, and cytoskeletal plasticity. The combination of multiphoton nanosurgery and *in vivo* imaging represents a promising tool for probing and disrupting neuronal circuits. The potential of using this precise optical method to perturb individual synapses cannot be overstated. Using multiphoton nanosurgery, the synaptic organization of the brain can now be teased apart *in vivo* to understand the microcircuitry of neuronal networks. Microscopy, therefore, has clearly expanded from its initial and very important role in pure imaging (yet improving impressively also in that area) to become a fully comprehensive and versatile tool in biological investigation.

Acknowledgments

We thank Rodney P. O'Connor and Francesco Vanzi for helpful discussions about the chapter.

References

1 Denk, W., Strickler, H.J., and Webb, W.W. (1990) Two-photon laser scanning fluorescence microscopy. *Science*, **248**, 73–76.

2 Zipfel, W.R., Williams, R.M., and Webb, W.W. (2003) Nonlinear magic: multiphoton microscopy in the biosciences. *Nat. Biotechnol.*, **21**, 1369–1377.

3 Helmchen, F. and Denk, W. (2005) Deep tissue two-photon microscopy. *Nat. Methods*, **2**, 932–940.

4 Svoboda, K. and Yasuda, R. (2006) Principles of two-photon excitation microscopy and its applications to neuroscience. *Neuron*, **50**, 823–839.

5 Konig, K., Riemann, I., and Fritzsche, W. (2001) Nanodissection of human chromosomes with near-infrared femtosecond laser pulses. *Opt. Lett.*, **26**, 819–821.

6 Sacconi, L., Tolic-Norrelykke, I.M., Antolini, R., and Pavone, F.S. (2005) Combined intracellular three-dimensional imaging and selective nanosurgery by a nonlinear microscope. *J. Biomed. Opt.*, **10**, 14002–14005.

7 Watanabe, W. and Arakawa, N. (2004) Femtosecond laser disruption of subcellular organelles in a living cell. *Opt. Express*, **12**, 4203–4213.

8 Supatto, W., Débarre, D., Moulia, B., Brouzés, E., Martin, J., Farge, E., and Beaurepaire, E. (2005) *In vivo* modulation of morphogenetic movements in Drosophila embryos with femtosecond laser pulses. *Proc. Natl. Acad. Sci. USA*, **102**, 1047–1052.

9 Nimmerjahn, A., Kirchhoff, F., and Helmchen, F. (2005) Resting microglial cells are highly dynamic surveillants of brain parenchyma *in vivo*. *Science*, **308**, 1314–1318.

10 Nishimura, N., Schaffer, C.B., Friedman, B., Tsai, P.S., Lyden, P.D., and Kleinfeld, D. (2006) Targeted insult to subsurface cortical blood vessels using ultrashort laser pulses: three models of stroke. *Nat. Methods*, **3**, 99–108.

11 Tsai, P.S., Friedman, B., Ifarraguerri, A.I., Thompson, B.D., Lev-Ram, V., Schaffer, C.B., Xiong, C., Tsien, R.Y., Squier, A., and Kleinfeld, D. (2003) All-optical histology using ultrashort laser pulses. *Neuron*, **39**, 27–41.

12 Yanik, M.F., Cinar, H., Cinar, H.N., Chisholm, A.D., Jin, Y., and Ben-Yakar, A. (2004) Neurosurgery: functional regeneration after laser axotomy. *Nature*, **432**, 822.

13 Chung, S.H., Clark, D.A., Gabel, C.V., Mazur, E., and Samuel, A.D. (2006) The role of the AFD neuron in C. elegans thermotaxis analyzed using femtosecond laser ablation. *BMC Neurosci.*, **7**, (http://www.biomedcentral.com/1471-2202/7/30).

14 Sacconi, L., O'Connor, R.P., Jasaitis, A., Masi, A., Buffelli, M., and Pavone, F.S. (2007) *In vivo* multiphoton nanosurgery on cortical neurons. *J. Biomed. Opt.*, **12**, 050502.

15 Vogel, A. and Venugopalan, V. (2003) Mechanisms of pulsed laser ablation of biological tissues. *Chem. Rev.*, **103**, 577–644.

16 Vogel, A., Noack, J., Hüttman, G., and Paltauf, G. (2005) Mechanisms of femtosecond laser nanosurgery of cells and tissues. *Appl. Phys. B*, **81**, 1015–1047.

17 Koenig, K., Riemann, I., Fischer, P., and Halbhuber, K.H. (1999) Intracellular nanosurgery with near infrared femtosecond laser pulses. *Cell Mol. Biol.*, **45**, 195–201.

18 Tirlapur, U.K. and Konig, K. (2002) Cell biology – targeted transfection by femtosecond laser. *Nature*, **418**, 290–291.

19 Zeira, E., Manevitch, A., Khatchatouriants, A., Pappo, O., Hyam, E., Darash-Yahana, M., Tavor, E., Honigman, A., Lewis, A., and Galun, E. (2003) Femtosecond infrared laser an efficient and safe in vivo gene delivery system for prolonged expression. *Mol. Ther.*, **8**, 342.

20 Smith, N.I., Fujita, K., Kaneko, T., Katoh, K., Nakamura, O., Kawata, S., and Takamatsu, T. (2001) Generation of calcium waves in living cells by pulsed-laser-induced photodisruption. *Appl. Phys. Lett.*, **79**, 1208.

21 Oehring, H., Riemann, I., Fischer, P., Halbhuber, K.J., and König, K. (2000) Ultrastructure and reproduction behaviour of single CHO-K1 cells exposed to near infrared femtosecond laser pulses. *Scanning*, **22**, 263.

22 Heisterkamp, A., Maxwell, I.Z., Mazur, E., Underwood, J.M., Nickerson, J.A., Kumar, S., and Ingber, D.E. (2005) Pulse energy dependence of subcellular dissection by femtosecond laser pulses. *Opt. Express*, **13**, 3690.

23 Shen, N., Datta, D., Schaffer, C.B., LeDuc, P., Ingber, D.E., and Mazur, E. (2005) Ablation of cytoskeletal filaments and mitochondria in cells using a femtosecond laser nanoscissor. *Mech. Chem. Biosystems*, **2**, 17–26.

24 Maxwell, I., Chung, S., and Mazur, E. (2005) Nanoprocessing of subcellular targets using femtosecond laser pulses. *Med. Laser Appl.*, **20**, 193–200.

25 Bourgeois, F. and Ben-Yakar, A. (2008) Femtosecond laser nanoaxotomy properties and their effect on axonal recovery in C. elegans. *Opt. Express*, **15**, 8521–8531.

26 Tsai, P.S., Blinder, P., Migliori, B.J., Neev, J., Jin, Y., Squier, J.A., Kleinfeld, D. (2009). Plasma-mediated ablation: An optical tool for submicrometer surgery on neuronal and vascular systems. *Curr. Opin. Biotech.*, **20**, 1–10.

27 Tolic-Norrelykke, I.M., Sacconi, L., Thon, G., and Pavone, F.S. (2004) Positioning and elongation of the fission yeast spindle by microtubule-based pushing. *Curr. Biol.*, **14**, 1181–1186.

28 Shimada, T., Watanabe, W., Matsunaga, S., Higashi, T., Ishii, H., Fukui, K., Isobe, K., and Itoh, K. (2005) Intracellular disruption of mitochondria in a living HeLa cell with a 76-MHz femtosecond laser oscillator. *Opt. Express*, **13**, 9869–9880.

29 Kumar, S., Maxwell, I.Z., Heisterkamp, A., Polte, T.R., Lele, T., Salanga, M., Mazur, E., and Ingber, D.E. (2006) Viscoelastic retraction of single living stress fibers and its impact on cell shape, cytoskeletal organization, and extracellular matrix mechanics. *Biophys. J.*, **90**, 3762–3773.

30 White, J., Southgate, E., Thomson, N., and Brenner, S. (1986) The structure of the Caenorhabditis elegans nervous system. *Philos. Trans. R. Soc. London (Biol.)*, **314**, 1–340.

31 Feng, G., Mellor, R.H., Bernstein, M., Keller-Peck, C., Nguyen, Q.T., Wallace, M., Nerbonne, J.M., Lichtman, J.W., and Sanes, J.R. (2000) Imaging neuronal subsets in transgenic mice expressing multiple spectral variants of GFP. *Neuron*, **28**, 41–51.

32 Isenberg, G., Bielser, W., Meier-Ruge, W., and Remy, E. (1976) Cell surgery by laser microdissection: a preparative method. *J. Microsc.*, **107**, 19–24.

33 Berns, M.W., Aist, J., Edwards, J., Strahs, K., Girton, J., McNeill, P., Rattner, J.B., Kitzes, M., Hammerwilson, M., Liaw, L.H., Siemens, A., Koonce, M., Peterson, S., Brenner, S., Burt, J., Walter, R., Bryant, P.J., Vandyk, D., Coulombe, J., Cahill, T., and Berns, G.S. (1981) Laser micro-surgery in cell and developmental biology. *Science*, **213**, 505–513.

34 Koonce, M.P., Strahs, K.R., and Berns, M.W. (1982) Repair of laser-severed stress fibers in myocardial non-muscle cells. *Exp. Cell Res.*, **141**, 375–384.

35 Strahs, K.R. and Berns, M.W. (1979) Laser microirradiation of stress fibers and intermediate filaments in non-muscle cells from cultured rat-heart. *Exp. Cell Res.*, **119**, 31–45.

36 Leslie, R.J. and Pickett-Heaps, J.D. (1983) Ultraviolet microbeam irradiations of mitotic diatoms: investigation of spindle elongation. *J. Cell Biol.*, **96**, 548–561.

37 Grill, S.W., Gonczy, P., Stelzer, E.H., and Hyman, A.A. (2001) Polarity controls forces governing asymmetric spindle positioning in the Caenorhabditis elegans embryo. *Nature*, **409**, 630–633.

38 Grill, S.W., Howard, J., Schaffer, E., Stelzer, E.H., and Hyman, A.A. (2003) The distribution of active force generators controls mitotic spindle position. *Science*, **301**, 518–521.

39 Khodjakov, A., Cole, R.W., and Rieder, C.L. (1997) A synergy of technologies: combining laser microsurgery with green fluorescent protein tagging. *Cell Motil. Cytoskeleton*, **38**, 311–317.

Part Two
Light–Molecule Interaction Mechanisms

3
Interaction of Pulsed Light with Molecules: Photochemical and Photophysical Effects

Gereon Hüttmann

3.1
Introduction

Radiation of the visible spectral range and the neighboring infrared and ultraviolet wavelengths interacts with the electrons of the outer shell of a large number of biologically relevant molecules. Three main processes are initiated when photons are absorbed by the electronic system of a molecule. First, excitation energy may be converted into heat. Second, a photon may be emitted (luminescence) and, third, photochemical reactions can follow. With continuous wave (CW) illumination photon fluxes are low and only one photon interacts with one ground state molecule. For CW-excitation, photophysical and often photochemical effects usually obey the Bunsen–Roscoe law and depend linearly on the total light dose.

Generally, pulsed irradiation allows significantly higher irradiances or photon flux density, because the photons are concentrated over a certain time period. Focusing by a high NA (numerical aperture) may further increase the irradiance. With femtosecond pulses, irradiances in the $TW\,cm^{-2}$ range or even higher are possible. Today the irradiances used in biomedical applications range from $10^{-3}\,cm^{-2}$ to $10^{15}\,W\,cm^{-2}$ (Figure 3.1). Two regimes may be discriminated. Below $10^{9}\,W\,cm^{-2}$, nonlinear effects of pulsed lasers are mainly caused by the interaction of photons with excited molecules. Consequently, excited state absorption and bleaching of the ground state may significantly change the photophysical behavior. The interaction cross-sections, and the kinetics of the involved states have a significant influence on the photophysical and photochemical behavior. At higher irradiances more than one photon can interact simultaneously with a molecule. In this way, molecular states are accessible that have a higher energy than that of a single photon.

To understand the processes involved we first discuss relevant photophysical properties of molecules with special emphasis on the dynamics between states. Sequencing the effects of ground state bleaching and excited state absorption will be discussed, which dominate the interaction in the kW and MW range. Multiphoton excitation via virtual states that are dominant in the GW and TW range are discussed in Section 3.4.

Laser Imaging and Manipulation in Cell Biology. Edited by Francesco S. Pavone
Copyright © 2010 WILEY-VCH Verlag GmbH & Co. KGaA, Weinheim
ISBN: 978-3-527-40929-7

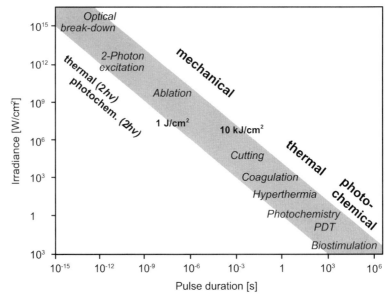

Figure 3.1 Irradiance typically accessible with different pulse widths, and the resulting interaction mechanisms with tissue.

As laser technology progresses, new ranges of pulse width and irradiance became available, and interest in nonlinear effects of pulsed irradiation on molecules has grown in recent years. Fluorescence microscopy and cell surgery using multiphoton excitation are probably the most prominent examples of new applications based on the interaction of ultrashort laser pulses with molecules. Section 3.5 covers the relevance of the interaction of pulsed light with molecules for biomedical imaging and optical manipulation of biomolecules and tissues.

3.2
Basic Photophysics

The interaction of molecules with pulsed light is determined by the photophysics of molecules, that is, the energy structure, the transitions possible between the energy levels, and the interaction mechanisms with light. This chapter presents the basics of photophysics that are necessary to understand the interaction of pulsed light with molecules.

In terms of classical physics, light is described by a superposition of waves that are characterized by a wavelength (λ), a frequency (f), and a polarization state. The wavelength and frequency are linked by the phase velocity (c):

$$f = \frac{c(\lambda)}{\lambda} \tag{3.1}$$

which in a vacuum is an universal constant, c_0, but in matter depends on the wavelength. The classical description is very successful in describing the macroscopic propagation of light. The most relevant effects like beam propagation, refraction, diffraction, and interference are readily understood in the framework of the classical theory of electromagnetic waves.

However, in its interaction with matter the wave model fails and light behaves more like particles that carry a momentum (p) and an energy (E), both of which are linked via Planck's constant (h) with the wavelength and the frequency, respectively:

$$p = h/\lambda$$
$$E = hf$$

(3.2)

According to the principles of physics, in the interaction of photons with matter both momentum and energy have to be preserved. In the world of atoms and molecules, described by quantum mechanics, an exchange of energy is only possible in discrete quantities, which are determined by internal discrete molecular states with distinct energies. Therefore, to understand the interaction of light with matter we have to discuss, first, structure of the molecular energy levels, which are in principle unique for different molecules but, in fact, follow a general scheme.

3.2.1
Electronic States of Molecules and the Jablonski Diagram

Quantum theory predicts that molecules possess discrete states that are characterized by certain quantum numbers, which in turn can be related to physical parameters of the electron system and the nuclei. For the simplest atomic system consisting of a proton and an electron, the hydrogen atom, the different quantum mechanical states, with their energy, their angular momentum, and symmetry of the associate quantum mechanical wavefunctions, can be calculated exactly. For larger atoms and molecules an exact calculation is not possible. Approximations have to be used, which neglect the electrons in filled inner shells and separate the motions of electrons and nuclei (justified by the large difference in their masses). Then the total energy of a molecular state is added up from the energy of the outer electrons and the vibrational and rotational modes. In addition, the spin of the electrons, which can take the two values ($+\frac{1}{2}\hbar$ and $-\frac{1}{2}\hbar$), results in a magnetic moment that interacts with the magnetic moment of the nucleus and modifies the electronic energy in the presence of a magnetic field. For the total energy of a molecular state only the sum of all electron spins matters. Usually, due to the Pauli principle, which allows at maximum two electrons with opposite spin to populate one electronic state, the total spin is zero for stable molecules in the ground state, because the spin of the two paired electrons in one electronic state compensates each other. Only radicals with an uneven number of electrons have the spin $+\frac{1}{2}\hbar$.

The contributions of the different molecular states to the total energy of a molecule varies from a few eV for the electronic states, over a few tens of electron volts (eV) for vibrations to a few tens of meV for molecular rotation over two orders. The states of a molecule can be ordered according to their energy, starting from the electronic

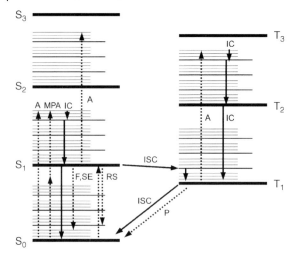

Figure 3.2 Energy diagram (Jablonski diagram) of the energy states of a molecule.

lowest energy state without vibration and rotation energy up to the ionization level. An energy ladder is obtained, with the electronic energy states intercalated by vibrational and rotational states (Figure 3.2). As soon as the molecule is not in the ground state there is the possibility of a total electron spin larger than zero. For example, two anti-parallel electrons in different electronic states will result in a total spin of $\hbar$. These states are called triplet states, since in contrast to the singlet states with a total spin of zero the triplet energy levels will split into three levels in an external magnetic field. Owing to the conversion of angular momentum a change between singlet and triplet states is usually a rare event. Therefore, in an energy diagram (called the Jablonski diagram) the energy ladders of singlet and triplet states are usually drawn separately (Figure 3.2).

Different states determine in their manifold the absorption spectra of the molecules. Larger molecules in fluid phase have continuous spectra because of the density of the vibrational and rotational energy levels and significant inhomogeneous broadening. The absorption strength, which is usually expressed as the absorption cross-section, is for strongly absorbing molecules of the order of 10^{-16} cm^2, that is, a few ångström squared (Å^2).

3.2.2
Changes between States

Changes between different states can either involve photons or can occur radiationlessly by exchanging the energy difference by heat. In transitions that involve photons, the energy difference is provided by the photon energy, which depends on the wavelength. Therefore, the interaction depends on the wavelength. The transition between the energy states can be spontaneous or can be triggered by a photon (stimulated processes):

Non-radiative transitions:

(i) Internal conversion (IC): spontaneous change between states *without* spin change; excess energy is converted into heat.

(ii) Intersystem crossing (ISC): spontaneous change between states *with* spin change; excess energy is converted into heat.

(iii) Thermal activation (TA): change between states; necessary excitation energy is provided by thermal energy.

Radiative transitions involving one photon:

(i) Absorption (A): change to a higher energy state is triggered by a photon; energy difference is provided by the photon.

(ii) Stimulated emission (SE): change to lower energy states is triggered by a photon; an identical photon is emitted.

(iii) Fluorescence (F): spontaneous change between states by emitting a photon *without* spin change.

(iv) Phosphorescence (P): spontaneous change between states emitting a photon *with* spin change.

(v) Raman scattering (RS): Change between vibrational triggered by a photon; energy difference is provided the change of the energy of the scattered photon.

Radiative transitions involving more than one photon:

(i) Multiphoton absorption (MPA): the energy of several photons is used for a change to a higher energy state.

(ii) Stimulated Raman scattering (SRS): Raman process, which is triggered by a second photon with shifted frequency.

(iii) Coherent Anti-Stokes Raman scattering (CARS): stimulated Raman process involving four photons, resulting in scattered photons with a higher energy.

At low irradiance only *one* photon will interact with *one* molecule, which is in the ground state. The photon energy and therefore the wavelength determine states of the singlet system (i.e., S_1, S_1 + vib. Energy, S_2) that can be accessed by absorption. After passing through different states the molecule will eventually return to the ground state without interaction with a second photon. The interaction will be a single photon/ground state interaction, as long as the average time between two interactions of the molecule with a photon is significantly longer than the deactivation time of the molecule. Pulsed radiation may change the photophysics, if the rate of absorption (k_A) from one of the involved states is high enough to compete with the typical relaxation rate (k) of one of the photophysical processes. Table 3.1 shows typical time constants ($\tau = 1/k$) for the photophysical processes discussed so far. The fastest spontaneous photophysical processes are radiationless changes within the singlet or triplet, that is, the relaxation of rotational and vibrational energy, or changes from higher excited electronic states to the S_1 or the T_1, which proceed in femto- or picoseconds ($k = 10^{11}-10^{14}$ s^{-1}). Deactivation between S_1 and S_0 by fluorescence and IC takes pico- to nanoseconds ($k = 10^7-10^{11}$ s^{-1}). Though connected with a spin change, in some molecules the ISC rate between S_1 and T_1 can be high enough to compete with the $S_1 \rightarrow S_0$ transition. Similar times are needed in certain molecules for ISC between the S_1

Table 3.1 Typical time constants for elementary photophysical processes.

Photophysical process	Involved states	Rate constant	Characteristic times (s)
Internal conversion	$S_i \rightarrow S_j + \text{heat}$		$10^{-14} - 10^{-11}$
	$T_i \rightarrow T_j + \text{heat}$		
	$S_i \rightarrow S_1 + \text{heat}$		$10^{-14} - 10^{-11}$
	$T_i \rightarrow T_1 + \text{heat}$		$10^{-14} - 10^{-11}$
	$S_1 \rightarrow S_0 + \text{heat}$		$10^{-11} - 10^{-7}$
Intersystem crossing	$S_1 \rightarrow T_j$		$10^{-11} - 10^{-8}$
	$T_1 \rightarrow S_0 + \text{heat}$		$10^{-5} - 10$
Linear absorption	$h\nu + S_0 \rightarrow S_j$	$B_{0j}/c = \sigma_A E_p$	10^{-15} (emission process)
	$h\nu + T_1 \rightarrow T_j$		
Stimulated emission	$h\nu + S_i \rightarrow S_0 + 2h\nu$	$B_{0j}/c = \sigma_{SE} E_p$	10^{-15} (emission process)
Fluorescence	$S_i \rightarrow S_0 + h\nu$	$\dfrac{8\pi h f^3}{c^3} B_{0j}$	$10^{-11} - 10^{-7}$
Phosphorescence	$T_1 \rightarrow S_0 + h\nu$		$10^{-11} - 10^{-7}$

and T_j. The longest time constants are observed for the change from T_1 into the ground state, which needs microseconds up to seconds because a spin change is necessary for this transition.

The kinetics of a change between the state i and state j are described mathematically by the rate W_{ij}, which quantifies the number events per time. For spontaneous events these rates are given by the population of a state N_i (number of molecules in state i) multiplied with a rate constant k_{ij}:

$$W_{ij} = k_{ij} N_i \qquad (3.3)$$

Absorption and stimulated emission need an additional photon. Therefore, the transition rates (W_{ij}) also depend on the volumetric photon density (ϱ). The rate constants for absorption and stimulated emission, which are also called the Einstein coefficient B_{ij}, are given by:

$$W_{ij} = B_{ij} N_i \varrho \qquad (3.4)$$

For two given states i and j the rate constants for absorption (B_{ij}) and stimulated emission (B_{ji}) are the same. The rate constant for the spontaneous emission of photons (fluorescence and phosphoresces) is often referred to as the Einstein coefficient A_{ij}:

$$A_{ij} = \frac{8\pi h f^3}{c^3} B_{ij} \qquad (3.5)$$

A_{ij} is closely linked to these rate constants because, ultimately, absorption, emission, and stimulated emission are all determined by the transition dipole moment (μ_{ij}) between the two the states, which can by calculated from a quantum mechanical description of the processes. A high absorption is also associated with high emission probabilities or short intrinsic fluorescence lifetime of the excited state.

The temporal dynamics of the changes between the states is mathematically described by of by rate equations and the rate constants of the different photophysical processes. For each state j a differential equation is formulated, which describes the change of the population $[(\partial/\partial t)[S_j]$ or $(\partial/\partial t)[T_j]]$ as a function of populations $([S_i]$ or $[T_i])$ of the different states using the rate constants (k_{ij}):

$$\frac{\partial}{\partial t}[S_j] = \sum_{i \neq j} k_{ij}[S_i] + \left(\sum_{i \neq j} k_{ij} \right) [S_j] \tag{3.6}$$

For absorption and stimulated emission the rate constants are actually determined by the Einstein coefficient, B_{ij} or A_{ij}, and the photon density (ϱ) or, synonymously, the cross-sections (σ_{ij}) multiplied with the photon flux density, $E_p = \varrho c$, or irradiance (E):

$$k_{ij} = \sigma_{ij} E_p = \frac{\sigma_{ij}}{hf} E = \frac{\varepsilon \log(10)}{hf} E \tag{3.7}$$

Instead of the absorption cross-section (σ) in practice, usually, the molar extinction coefficient (ε) is used to quantify the absorption of a molecule.

3.3
Bleaching and Excited State Absorption

New effects of pulsed irradiation will arise when the excitation rate begins to compete with internal photophysical processes in the molecules. For a typical molecule with a molar extinction coefficient of the order of $10^5 \, \mathrm{mol \, l^{-1} \, cm^{-1}}$ and an irradiance of the order of $\mathrm{W \, cm^{-2}}$ the excitation rate is only $10^3 \, \mathrm{s^{-1}}$. This is lower than the rates of all photophysical processes. Even the $T_1 \rightarrow S_0$ transition is usually much faster $(10^6 \, \mathrm{s^{-1}})$ because of the quenching of the triplet energy by oxygen or other substances. However, with irradiances of $\mathrm{kW \, cm^{-2}}$ the excitation rate starts to compete with the triplet decay rate (Figure 3.3). These irradiances are typically

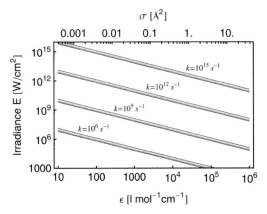

Figure 3.3 Irradiance (E) necessary for a certain excitation rate (k) as a function of molar absorption coefficient (ε) or absorption cross-section (σ). Calculation was carried out for the wavelengths 400 nm, 525, 650, and 800 nm, which gives slightly shifted lines.

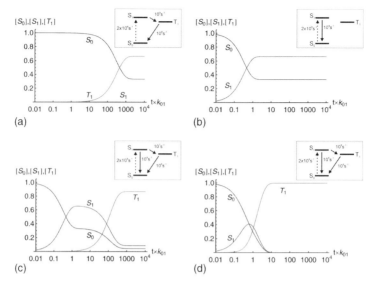

Figure 3.4 (a–d) Change of the population in a three-level system (S_0, S_1, T_1) under continuous irradiation. (a) Excitation rate is twice the deactivation rate of the triplet state, which is one thousand times slower than the transition $S_0 \rightarrow S_1$; (b) excitation rate is increased to twice the transition rate $S_1 \rightarrow S_0$; the triplet quantum yield is 0; (c) excitation rate twice the transition rate $S_1 \rightarrow S_0$; triplet yield is 0.01; (d) excitation rate twice the transition rate $S_1 \rightarrow S_0$; triplet yield is 0.5.

produced by focused CW laser, or pulses with microsecond duration. Already, microwatts focused on an area with a diameter of 0.5 μm yield an irradiance in the kilowatt range. Irradiation extended over microseconds may lead to bleaching of the ground state (Figure 3.4a). Depending on the absorption spectrum of T_1 an excitation to higher triplet states (triplet–triplet absorption) may additionally occur.

For an absorption coefficient of 10^5 mol^{-1} cm^{-1}, the excitation rate of a 1 mJ cm^{-2} nanosecond pulse or a few milliwatts focused on 0.5 μm can compete with deactivation of the S_1 state (Table 3.2). Saturation can already happen in the S_1 state. From here, additionally, the T_1 will be populated, trapping also molecules in the triplet state. For a simplified system that consists of only the three states S_0, S_1, and T_1 the

Table 3.2 Irradiance, excitation rates and competing photophysical processes at a radiant exposure of 1 mJ cm^{-1} but different pulse widths.

Pulse width	Irradiance @ 1 mJ cm^{-2}	Excitation rate @ $\varepsilon = 10^5$ l mol^{-1}cm^{-1} (s^{-1})	Competing processes
1 ms	1 W cm^{-2}	10^3	
1 μs	1 kW cm^{-2}	10^6	ISC : $T_1 \rightarrow S_0$
1 ns	1 MW cm^{-2}	10^9	Fluor., IC : $S_1 \rightarrow S_0$
1 ps	1 GW cm^{-2}	10^{12}	IC : $S_n \rightarrow S_1, T_n \rightarrow T_1$
1 fs	1 TW cm^{-2}	10^{15}	Absorption process

following rate equation can be derived from Equation (3.6):

$$\frac{\partial}{\partial t}[S_0] = -\sigma E_p[S_0] + k_{S1 \to S0}[S_1] + k_{T1 \to S0}[T_1]$$

$$\frac{\partial}{\partial t}[S_1] = \sigma E_p[S_0] - (k_{S1 \to S0} + k_{ISC})[S_1] \qquad (3.8)$$

$$\frac{\partial}{\partial t}[T_1] = k_{ISC}[S_1] - k_{T1 \to S0}[T_1]$$

The rate constants for the spontaneous processes can be expressed by the lifetimes, τ_S and τ_T, of the first excited singlet and triplet state and the triplet quantum yield:

$$\tau_S = 1/(k_{S1 \to S0} + k_{ISC}); \quad \tau_T = 1/k_{T1 \to S0}$$
$$\eta_T = k_{ISC}/(k_{ISC} + k_{S1 \to S0}) \qquad (3.9)$$

Figure 3.4a–d shows the development of the population, under continuous irradiation, calculated for different excitation rates and triplet yields. The transition $S_1 \to S_0$ was assumed to be a thousand times faster than the deactivation rate of the triplet state (e.g. 10^9 s^{-1} for $S_1 \to S_0$ and 10^6 s^{-1} for $T_1 \to S_0$). With a unity quantum yield for triplet formation even at an excitation rate twice the deactivation rate of the triplet (2×10^6 s^{-1}), which corresponds for a high absorbing molecule to an irradiance in the kW cm^{-2} range, the T_1 population increases until at the steady-state two-thirds of the molecules will be in the triplet state (Figure 3.4a). Excitation rates of 2×10^9 s^{-1} (irradiation in the MW cm^{-2}), which is twice the rate of $S_1 \to S_0$, are simulated in Figure 3.4b–d. If the triplet yield is zero, accumulation in S_1 is observed, which also reaches in the steady-state regime 66% population (Figure 3.4b). If there is a certain probability for intersystem crossing, accumulation in the S_1 state will only be transient and eventually all molecules will end up in the triplet state (Figure 3.4c, d).

As soon as a significant number of molecules are in the excited state, excited state absorption gives access to higher electronic states. From here thermalization of the additionally absorbed energy by rapid internal conversion will result. In addition, bleaching and photochemistry from the higher energy levels may be enhanced. The rate for excited state absorption depends directly on the irradiance, absorption cross-section, and population of the excited state, which itself depends on the irradiance.

For modeling the excited state dynamics the rate constants k_{ij} have to be known. Different experimental methods exist to determine the rate constants. Indirectly, the dependence of a photophysical or photochemical response on the irradiance can be measured and the measured data can be fitted to model, which incorporates the unknown constants as free parameters. Possible measurable parameters are the total absorption, fluorescence intensity, or the yield of the photochemical reaction [1, 2]. The model, to which the data are fitted, has to include the photophysics and the measurement geometry (e.g., inhomogeneity of the irradiation). The simpler the model and the fewer parameters are unknown the better the result will be. More direct information on the rate constants will be obtained by a measurement of the excited state life times and the absorption cross-sections. Fluorescence lifetime measurements give, at low irradiation, directly the sum of rate constants of the spontaneous processes starting from the S_1 $(k_{S_1 \rightarrow S_0} + k_{S_1 \rightarrow T_1})$. Time-resolved absorption measurements after excitation provide information on cross-sections of the S_1, T_1 life-time, and the cross-sections for the $S_1 \rightarrow S_n$ and $T_1 \rightarrow S_n$ transitions.

3.4
Multiphoton Absorption and Ionization

Irradiances in the $GW\,cm^{-2}$ to $TW\,cm^{-2}$ range result in excitation rates that compete with internal conversion and also with the time for the absorption process itself. A few milliwatts of a 100-femtosecond pulses train with $80\,MHz$ repetition rate focused on $0.5\,\mu m$ results in hundreds of $GW\,cm^{-2}$. For a high absorbing transition the excitation rates reach $10^{15}\,s^{-1}$, which is the reciprocal of the duration of the absorption event. Consequently, more than one photon may take part in the absorption process. Maria Göppert-Meyer showed in 1931, by quantum mechanical calculations, that two-photon absorption processes are indeed possible [3]. Thirty years later two-photon excited fluorescence was demonstrated [4]. At irradiances of $100\,GW\,cm^{-2}$, two-photon absorption in organic dyes is observed [5]. At higher irradiances, three-photon absorption [6] – and in the $TW\,cm^{-2}$ range even multi-photon ionization [7] – may occur.

In multiphoton processes, the excitation rate is proportional to the average irradiance ($\langle E \rangle$) raised to the power of the number of involved photons [8]:

$$k_n = g_n \sigma_n \left(\frac{\langle E \rangle}{hc/\lambda} \right)^n$$

with

$$g_n = \frac{\int E^n \, dt}{\left(\int E \, dt \right)^n}$$

(3.10)

The factor g_n accounts for the reduction of effective interaction time with a laser pulse due to the nonlinear response of the medium. The cross section (σ_n) decreases strongly with the number (n) of simultaneously absorbed photons. Typical cross-sections for two- and three-photon absorption are listed below [5]:

- $\sigma_1 = 10^{-17}$ to $10^{-15}\,cm^2$ (corresponding to $\varepsilon = 2.6 \times 10^3$ to 2.6×10^5 l mol cm),
- $\sigma_2 = 10^{-51}$ to $10^{-48}\,cm^4$ s (e.g., the flavin FMN: $2 \times 10^{-51}\,cm^4$ s),
- $\sigma_3 = 10^{-83}$ to $10^{-82}\,cm^6\,s^2$.

Because of the low absorption cross-sections and the dependence on the square or cube of the irradiance, the photon densities are only high enough in strongly focused beams to allow significant multiphoton absorption.

To make practical use of multiphoton absorption, repetitive femto- or picosecond mode-locked lasers are usually employed. For these systems the average excitation rate $(\bar{k}_{ij}^{(n)})$ depends then on the repetition rate (f), the pulse width (τ), the average radiant power (Φ), and the numerical aperture (NA), by which the beam is focused [8]:

$$\bar{k}_{ij}^{(n)} = \frac{g_n}{(\tau f)^{n-1}} \left(\frac{\Phi \, \pi \, NA^2}{h \, c_0 \, \lambda} \right)^n \sigma_{ij}^{(n)} \qquad (3.11)$$

By using a pulse train instead of CW irradiation the average excitation rate increases by $1/(\tau f)^{n-1}$. For a typical mode-locked Ti:sapphire femtosecond laser ($f = 80\,MHz$ and $\tau = 10\,fs$) the two-photon excitation probability is increased 125 000-fold. A three-photon process will benefit from a 10^{10}-fold increase.

Confinement of the interaction between light and biomolecules to a micrometer-sized spatial volume by multiphoton absorption finds applications in fluorescence microscopy [9, 10], spatially confined photochemistry [11, 12], and cell surgery [13]. Especially, multiphoton excited fluorescence microscopy of scattering tissues is today widely used [14]. In contrast to confocal microscopy, the longer possible excitation wavelengths are less scattered by the tissue, non-ballistic (i.e., scattered) fluorescence photons can additionally be detected, and photobleaching is limited to the focal area. Three-photon excitation is useful for fluorescence microscopy with UV absorption dyes [5], avoiding the use of visible or UV light for excitation. With even more photons absorbed, quasi-free electrons may be produced. These electrons can react with biomolecules or provide the starting electrons for plasma formation and optical breakdown [7].

3.5
Relevance for Biomedical Applications

The interaction of pulsed irradiation with molecules is important for various biomedical applications in which lasers or optical radiation is used. High excitation rates and multiphoton absorption, which are associated with pulsed irradiation, can reduce, on one hand, the efficacy of useful processes (e.g., fluorescence emission, photochemistry) and increase the bleaching of the chromophores. These effects are relevant at phototherapies (e. g. photodynamic therapy), laser scan microscopy, or single molecule biomedical analysis (e.g., fluorescence correlation spectroscopy). On the other hand, novel biophotonic applications like two-photon and multiphoton microscopy or laser cell surgery rely on the use of pulsed irradiation.

3.5.1
Effectiveness of Pulsed Lasers for Photodynamic Therapy

Photodynamic therapy (PDT) is a new clinical treatment modality, which uses light, oxygen, and a dye, that selectively accumulates in certain tissue structures to destroy precancerous lesions, tumors, or new proliferating vessels [15–17]. Although in certain cases radical formation may be involved [18], it is widely believed that the phototoxic action of PDT is mediated by a sensitization of molecular oxygen. By energy transfer from the dye, which is also called the sensitizer, molecular oxygen changes from the ground state to the first excited state. As a very unusual case, the ground state of oxygen is a triplet state and the first excited state is a highly reactive singlet state [19]. The change of the spin of oxygen has to be compensated by a change of the sensitizer multiplicity. Therefore, energy transfer is only possible from the sensitizer triplet state.

Photodynamic therapy relies on a photochemical reaction and the phototoxic effect should depend – as for other photochemical effects – only on the radiant exposure and not on the irradiance. For the sensitizer hematoporphyrin derivative (HPD) this behavior was shown in a range between 5 and $230\,\mathrm{mW\,cm^{-2}}$ *in vitro* [20, 21]. However, upon pulsed irradiation a reduction of the phototoxic effect was also observed [22, 23].

A reduction of the PDT efficacy may be expected for two reasons. First, oxygen depletion in the irradiated volume results if the photooxidation consumes more oxygen than can be delivered by circulation and diffusion. Second, transient bleaching by trapping of the molecules in the S_1 and T_1 state can reduce the effective absorption. Owing to the long time constants for diffusion, the oxygen depletion depends effectively on the average irradiance and the local oxygen supply. In contrast, transient bleaching is determined by the peak irradiance, the pulse length, and repetition rate. The effects of transient bleaching can be described quantitatively by a relatively simple model. Relevant energy levels are the S_0, S_1, and the T_1 state of the sensitizer and the T_0 and the S_1 of the oxygen molecule (Figure 3.5).

From the energy scheme the kinetic equations for the population of the involved states are readily derived:

$$\frac{\partial}{\partial t}[S_0] = -\sigma E_p[S_0] + (\sigma E_p + k_{IC})[S_1] + (k_\Delta + k_T)[T_1]$$

$$\frac{\partial}{\partial t}[S_1] = \sigma E_p[S_0] - (\sigma E_p + k_{IC} + k_{ISC})[S_1]$$

$$\frac{\partial}{\partial t}[T_1] = k_{ISC}[S_1] - (k_\Delta + k_T)[T_1]$$

$$\frac{\partial}{\partial t}[^1O_2] = k_\Delta[T_1]$$

(3.12)

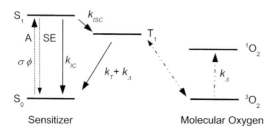

Figure 3.5 Relevant energy levels for the photophysics of photodynamic therapy.

It is advantageous to relate the rate constants to more direct measurable photophysical quantities such as the lifetimes (τ_{S1}, τ_{T1}) of the singlet and triplet states S_1 and T_1 and the quantum yields (η_T, η_Δ) for triplet formation and singlet oxygen production:

$$\tau_S = 1/(k_S + k_{ISC}); \qquad \tau_T = 1/(k_T + k_\Delta)$$
$$\eta_T = k_{ISC}/(k_{ISC} + k_{IC}); \quad \eta_\Delta = \eta_T\, k_\Delta/(k_\Delta + k_T) \tag{3.13}$$

The PDT efficacy (η_{PDT}) of the pulsed irradiation with respect to a CW irradiation is quantified by the ratio of singlet oxygen produced in both modalities at the same radiant exposure (H):

$$\eta_{PDT} = \frac{\int [^1O_2]dt}{\int \eta_\Delta\, \sigma\, E_p\, dt} = \frac{\int [^1O_2]dt}{\eta_\Delta\, \sigma H/hf} \tag{3.14}$$

Instead of using the quite complicated exact solution of the differential equation system an approximation can be used that assumes repetitive pulses with a low average irradiance and a separation of the pulses longer than the triplet lifetime (τ_{T1}). Reduced PDT efficacy results if the irradiance is high enough to capture the molecule in either the T_1- or the S_1-state. Under realistic conditions the former may happen with microsecond pulse whereas for the later nanosecond or shorter pulses are required. With this simplifying assumption for both cases, microsecond and nanosecond irradiation, simple formulas for the PDT efficacy (η_{PDT}) can be derived [24]:

$$\mu s\text{-pulses}: \quad \eta_{PDT} = \frac{1 - \exp(-\eta_T Y)}{\eta_T Y} \tag{3.15a}$$

$$ns\text{-pulses}: \quad \eta_{PDT} = \frac{1 - \exp(-2Y)}{2Y} \tag{3.15b}$$

with $Y = 1.92 \cdot 10^5\, \frac{mol\, cm^3\, nm}{l\, j}\, \varepsilon \cdot \lambda \cdot H$.

In this approximation the efficacy depends only on the radiant exposure per pulse, the molar extinction coefficient (ε), and for μs-pulses, when the excited molecules are accumulated in triplet state, on the triplet yield (η_T).

The model was verified experimentally *in vitro* by irradiating with the photosensitizer HPD incubated cells with microsecond pulses from a flashlamp-pumped dye laser. Different radiant exposures per pulse were used (Figure 3.6). Experimental results of other groups that compared pulsed and CW lasers for PDT also fit our model, though a quantitative comparison was not really possible because of a lack of exact photophysical and experimental data (Figure 3.7).

3.5.2
Reduction of Photobleaching in Laser Scanning Microscopy

Confocal laser scanning fluorescence microscopy has found wide applications in biomedical research for 3D imaging of biological samples [25]. For high sensitivity

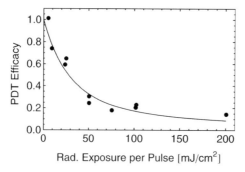

Figure 3.6 PDT Efficacy of microsecond irradiation on cells, which were sensitized by the photosensitizer HPD. The radiant exposure per pulse was varied between 10 and 200 mJ cm^{-2}.

and/or high imaging speed, irradiances in the kW cm^{-2} to MW cm^{-2} range are used. The maximal fluorescence rate of each molecule is limited by saturation and transient bleaching. During the lifetime of the excited state the molecule can emit only one photon. In addition, laser scanning and conventional fluorescence microscopy are hampered by photobleaching of the excited fluorophores. Especially for time-lapse imaging or the detection of low concentrations of fluorescence markers, photodamage of the fluorescing molecules (photobleaching) poses a severe limitation [26, 27]. Photobleaching may proceed via different pathways and in general follows quite complicated kinetics. For example, it has been shown that only a five-level molecular state model can describe the dependence of the bleaching of Rhodamine 6G on the irradiance [28]. Two dark states, a triplet state, and a radical ion state were identified. Efficient photobleaching after further excitation of these dark states was observed. An extended study of Rhodamine 6G was able to model the photobleaching kinetics by a simplified model that assumed steady state populations and a quasi-continuous irradiation [29]. At higher irradiance the fluorescence was

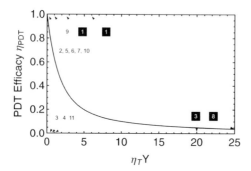

Figure 3.7 Comparison of experimental results for pulsed irradiation with the parameter $\eta_T Y$. Black numbers on white background indicate published experiments, in which the efficacy was not reduced under pulsed irradiation. White numbers on a black background refer to work in which a marked reduction was observed.

limited by photolysis from higher singlet and triplet states. Three regimes were observed: a linear increase of the fluorescence rate with irradiance as expected for low irradiance, saturation at higher irradiance, and even reduced fluorescence above $100 \, kW \, cm^{-2}$. Continuous and picosecond irradiation showed similar photobleaching behavior, whereas excitation with femtosecond pulses results in significant excitation of higher singlet states with negligible population of the triplet state.

The relevance of dark states for photobleaching was also demonstrated with repetitive excitation with picosecond and femtosecond pulses [30]. By varying the pulse separation from 25 ns to 2 μs the total amount fluorescence that could be obtained from the dye Atto532 and the fluorescing protein GFP was increased 4-fold and 20-fold, respectively. From the dependence of the increase of the fluorescence yield on the pulse separation it was concluded that a state with 1 μs lifetime must be involved in the photodestruction of the fluorescing molecules. Probably, the T_1 state is excited to higher unstable triplet states. Within one femto- or picosecond pulse the population of the T_1 is negligible, because it takes at least nanoseconds for a population of the T_1 by ISC to accrue. If at the arrival of the next pulse the triplet is not deactivated triplet–triplet absorption to unstable or photoreactive states may destroy the molecule. Triplet–triplet absorption can also destroy the molecule under CW irradiation if the excitation rate is higher than the triplet de-excitation rate (typically $10^6 \, s^{-1}$). Compared to a CW irradiation of $200 \, kW \, cm^{-2}$, a pulsed irradiation with 2 μs pulse separation gave a fivefold increase in yield of fluorescence photons for GFP and an eightfold increase for Atto532. The irradiance excited in both cases was approximately 5×10^7 molecules per second, that is, every 20 ns a photon was absorbed. Even with an order of magnitude lower absorption cross-section for triplet–triplet absorption the triplet absorption rate was high enough to compete with the triplet deactivation.

3.5.3
Super-Resolution by Optical Depletion of the Fluorescent State

Diffraction limits the minimal spot size by which light can be focused. Since Abbe it was widely believed that this is also the practical the resolution limit for microscopy. This diffraction limited resolution is usually expressed by the Rayleigh criterion for the minimal distance (d) of two resolvable points:

$$d = 0.61 \frac{\lambda}{NA} \tag{3.16}$$

Without aberrations, the diffraction limited resolution depends only on the wavelength (λ) and the numerical aperture (NA) of the focusing optics. This holds true only for linear imaging processes, as a nonlinear interaction of light with the object narrows the point spread function considerably. A strong optical nonlinearity can be introduced in the imaging process by the interaction with pulsed light. Stefan Hell proposed the use of stimulated emission in fluorescence imaging to narrow the emitting volume considerably [31]. Molecules have to be excited to a higher short-lived energy state (vibrational excited S_1^* or S_2), from which the molecule relaxes to

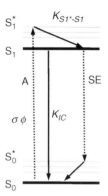

Figure 3.8 Principle of stimulated emission depletion (STED).

the lowest S_1 level. With a second coaxial beam, which has an annular intensity distribution, the molecules are quenched via stimulated emission. This is possible without re-excitation if a wavelength at the red edge of the emission spectrum is used, which leaves the molecule in a vibrationally excited state (S_0^*) (Figure 3.8). The stimulated emission process results in a highly nonlinear dependence of the fluorescence intensity on the radiant exposure of the stimulated emission depletion (STED) pulse and the fluorescence is quenched in the focal region by the ring shaped STED beam everywhere except for the optical axis. The zero irradiance region is mathematically a point. Therefore, by increasing the energy of the STED pulses, the fluorescing area can be made arbitrarily small, that is, the resolution can be increased without fundamental limit. For optimal resolution, pulse length and timing of the pulses should be carefully chosen. On one hand, the pulse length should be short enough so that excitation and stimulated emission rates are much faster than the fluorescence rate; on the other hand, the risk of photodamage by excited state absorption should be reduced. Typically for STED a peak irradiance of a few hundred MW cm^{-2} to some GW cm^{-2} is used [32]. Femtosecond pulses have been shown to increase photodamage due to multiphoton absorption. Nanosecond or longer pulses need increased average irradiance and carry the risk of re-exciting the possibly forming triplet state. Hence a pulse width below 100 ps is commonly used for excitation and a few hundred picoseconds for the stimulated emission depletion. The pulses should be separated by the relaxation time of the transition $S_1^* \rightarrow S_1$, which is a few picoseconds. Recently it was shown that STED microscopy is also possible with CW irradiation [33]. However, to be efficient, CW STED needs high excitation rates and is very sensitive to triplet formation. Fast scanning may reduce the effective exposure time and the build-up of a triplet population in order to reduce photodamage via the triplet states.

Beside depletion of the fluorescent state by stimulated emission, it is also possible to use excitation into a dark state, for example, the triplet state or radical anion, as a nonlinear saturable effect [34, 35]. In this case the necessary excitation has to compete only with the back reaction to S_1, which is significantly slower than the deactivation rate for the fluorescent S_1 state and the necessary irradiances can be reduced considerably.

3.6
Conclusions

The interaction of pulsed irradiation with molecules may lead to complex changes of the photophysics (fluorescence, thermalization of the excitation) and photochemistry. In contrast to low irradiance CW-irradiation, where the effect is determined only by the ground state absorption and the radiant exposure, the effect of pulsed irradiation additionally depends on the irradiance, irradiation time, pulse separation, and absorption cross-sections and lifetimes of the excited states. The effects of pulsed irradiation in biomedical applications are diverse, including reduced efficacy of fluorescence formation and photochemistry, increased photobleaching, and the excitation of higher energy levels by simultaneous absorption of more than one photon. As a nonlinear effect multiphoton absorption confines the interaction of light with molecules to the focal volume and has found considerable applications in microscopy and cell surgery.

Despite the complexity of excited state photophysics the basic processes relevant for biomedical applications are bleaching of the ground state absorption, stimulated emission from the S_1, excited state absorption from the S_1 and T_1 states, and multiphoton absorption. Numerous experimental studies suggest that excited state absorption reduces fluorescence yield and increases photobleaching. Dramatic changes in the photophysics and photochemistry are achieved by tailoring the time course of irradiation to the photophysical properties of the chromophore. Successful strategies for reduced photodamage and increased total fluorescence output per molecule have been developed.

STED (stimulated emission depletion), a new microscopic imaging strategy for increasing the resolution beyond the Abbe limit, is based on a sophisticated use of pulsed laser interaction.

Multiphoton fluorescence imaging, which uses direct excitation with two or more photons, is now widely used in microcopy. Though, in thick samples, it profits from the restriction of the excited volume to the thin layer that is actually imaged, photodamage in the focal volume still severely limits imaging speed and/or the length of image sequences in time-lapse microscopy. Multiphoton imaging uses GW cm^{-2} to TW cm^{-2} irradiance at a wavelength at which the tissue has no ground state absorption. After excitation to the S_1 or T_1 state, even at low populations or low absorption cross-sections a significant excited state absorption rate, which is associated with photodamage, is expected.

Adapting the irradiation pattern to chromophores is used to avoid excited state absorption and should reduce photodamage. The continuous development of laser sources and fast scanning devices will give access to a wider range of wavelengths, pulse width, and repetition rate. Together with a growing understanding of the excited state photophysics a reduction of photodamage in microscopy is expected.

References

1. Stiel, H., Teuchner, K., Leupold, D.,
Oberländer, S., Ehlert, J., and Jahnke, R.
(1991) Computer aided laser
spectroscopic characterization and
handling of molecular excited states.
Intell. Instrum. Comput., **9**, 79–88.

2. Stiel, H., Teuchner, K., Paul, A., Leupold,
D., and Kochevar, I.E. (1996) Quantitative
comparison of excited state properties
and intensity-dependent
photosensitization by Rose bengal. *J.
Photochem. Photobiol. B, Biol.*, **33** (3),
245–254.

3. Göppert-Mayer, M. (1931)
Über Elementarakte mit zwei
Quantensprüngen. *Ann. Phys.*, **9**,
273–294.

4. Kaiser, W. and Garrett, C.G.B. (1961)
Two-photon excitation in CaF2: Eu2 + .
Phys. Rev. Lett., **7** (6), 229.

5. Xu, C., Zipfel, W., Shear, J.B., Williams,
R.M., and Webb, W.W. (1996)
Multiphoton fluorescence excitation: new
spectral windows for biological nonlinear
microscopy. *Proc. Natl. Acad. Sci. USA*, **93**
(20), 10763–10768.

6. Szmacinski, H., Gryczynski, I., and
Lakowicz, J.R. (1996) Three-photon
induced fluorescence of the calcium
probe Indo-1. *Biophys. J.*, **70** (1) 547–555.

7. Vogel, A., Noack, J., Hüttmann, G., and
Paltauf, G. (2005) Mechanisms of
femtosecond laser nanosurgery of cells
and tissues. *Appl. Phys. B: Photophys. Laser
Chem.*, **81** (8), 1015–1047.

8. Koester, H.J., Baur, D., Uhl, R., and Hell,
S.W. (1999) Ca2 + fluorescence imaging
with pico- and femtosecond two-photon
excitation: signal and photodamage.
Biophys. J., **77** (4), 2226–2236.

9. Denk, W., Strickler, J.H., and Webb, W.W.
(1990) Two-photon laser scanning
fluorescence microscopy. *Science*, **248**
(4951), 73–76.

10. Zipfel, W.R., Williams, R.M., and
Webb, W.W. (2003) Nonlinear magic:
multiphoton microscopy in the
biosciences. *Nat. Biotechnol.*, **21** (11),
1369–1377.

11. Ogawa, K., Kobuke, Y. (2008) Recent
advances in two-photonn photodynamic

therapy. *Anticancer. Agents. Med. Chem.* **8**
(3), 269–279.

12. Li, L., Gattass, RR., Gershgoren, E.,
Hwang, H., Fourkas, J.T. (2009)
Achieving lambda/20 resolution by
one-color initiation and deactivation of
polymerization. *Science* **324** (5929),
910–913.

13. Vogel, A., Noaok, J., Hüitlmann, G.,
Paltauf, G. (2006) Femtosecond plasma-
mediated nanosurgery of cells and
tissues. In: *Laser Ablation* (C. Phipps C ed)
Springer, Heidelberg, pp 217–262.

14. Helmchen, F. and Denk, W. (2005) Deep
tissue two-photon microscopy.
Nat. Methods, **2** (12), 932–940.

15. Fisher, A.M., Murphree, A.L., and Gomer,
C.J. (1995) Clinical and preclinical
photodynamic therapy. *Lasers Surg. Med.*,
17 (1), 2–31.

16. Dougherty, T.J., Gomer, C.J., Henderson,
B.W., Jori, G., Kessel, D., Korbelik, M.,
Moan, J., and Peng, Q. (1998)
Photodynamic therapy. *J. Natl. Cancer
Inst.*, **90** (12), 889–905.

17. Robertson, C.A., Evans, D.H., and
Abrahamse, H. (2009) Photodynamic
therapy (PDT): a short review on cellular
mechanisms and cancer research
applications for PDT. *J. Photochem.
Photobiol. B, Biol.*, **96** (1), 1–8.

18. Allen, M.T., Lynch, M., Lagos, A.,
Redmond, R.W., and Kochevar, I.E. (1991)
A wavelength dependent mechanism
for Rose bengal-sensitized
photoinhibition of red cell
acetylcholinesterase. *Biochim. Biophys.
Acta.*, **1075** (1), 42–49.

19. Foote, C.S. (1976) Photosensitized
oxidation and singlet oxygen:
consequences in biological systems, in
Free Radicals in Biology (ed. W.A. Pryor),
Academic Press, New York, pp. 85–133.

20. Gibson, S.L. and Hilf, R. (1985)
Interdependence of fluence, drug dose
and oxygen on hematoporphyrin
derivative induced photosensitization of
tumor mitochondria. *Photochem.
Photobiol.*, **42** (4), 367–373.

21. Gomer, C.J., Rucker, N., Razum, N.J., and
Murphree, A.L. (1985) *In vitro* and *in vivo*

light dose rate effects related to hematoporphyrin derivative photodynamic therapy. *Cancer Res.*, **45** (5), 1973–1977.

22. Bellnier, D.A., Lin, C.W., Parrish, J.A., and Mock, P.C. (1984) Hemato-porphyrin derivative and pulsed laser photoradiation, in *Porphyrin Localization and Treatment of Tumors* (eds D.R. Doiron and C.J. Gomer), Alan R. Liss Inc., New York.

23. Barr, H., Boulos, P., Macrobert, A., Tralau, C., Phillips, D., and Bown, S. (1989) Comparison of lasers for photodynamic therapy with a phthalocyanine photosensitizer. *Lasers Med. Sci.*, **4** (1), 7–12.

24. Hüttmann, G., Heck, A., and Diddens, H. (1994) A simple model for the effectiveness of PDT with pulses laser sources. *Proc. SPIE*, **2134**, 223–235.

25. Pawley, J.B. (ed.) (2006) *Handbook of Biological Confocal Microscopy*, 3rd edn, SpringerScience + Business Media LCC, New York.

26. Tsien, R.Y., Ernst, L., and Waggoner, A. (2006) Fluorophores for confocal microscopy: photophysics and photochemistry, in *Handbook of Biological Confocal Microscopy*, 3rd edn (ed. J.B. Pawley), SpringerScience + Business Media, New York, pp. 338–352.

27. Eggeling, C., Widengren, J., Rigler, R., and Seidel, C.A.M. (1998) Photostability of fluorescent dyes for single-molecule spectroscopy: mechanisms and experimental methods for estimating photobleaching in aqueous solution, in *Applied Fluorescence in Chemistry, Biology and Medicine* (eds W. Rettig, B. Strehmel, and S. Schrader), Springer, New York, Berlin, pp. 193–240.

28. Eggeling, C., Widengren, J., Rigler, R., and Seidel, C.A.M. (1998) Photobleaching of fluorescent dyes under conditions used for single-molecule detection: evidence of two-step photolysis. *Anal. Chem.*, **70** (13), 2651–2659.

29. Eggeling, C., Volkmer, A., and Seidel, C.A. (2005) Molecular photobleaching kinetics of Rhodamine 6G by one- and two-photon induced confocal fluorescence microscopy. *ChemPhysChem*, **6** (5), 791–804.

30. Donnert, G., Eggeling, C., and Hell, S.W. (2007) Major signal increase in fluorescence microscopy through dark-state relaxation. *Nat. Methods*, **4** (1), 81–86.

31. Hell, S.W. and Wichmann, J. (1994) Breaking the diffraction resolution limit by stimulated emission: stimulated-emission-depletion fluorescence microscopy. *Opt. Lett.*, **19** (11), 780–782.

32. Harke, B., Keller, J., Ullal, C.K., Westphal, V., Schonle, A., and Hell, S.W. (2008) Resolution scaling in STED microscopy. *Opt. Express*, **16** (6), 4154–4162.

33. Willig, K.I., Harke, B., Medda, R., and Hell, S.W. (2007) STED microscopy with continuous wave beams. *Nat. Methods*, **4** (11), 915–918.

34. Heintzmann, R., Jovin, T.M., and Cremer, C. (2002) Saturated patterned excitation microscopy–a concept for optical resolution improvement. *J. Opt. Soc. Am. A*, **19** (8), 1599–1609.

35. Folling, J., Bossi, M., Bock, H., Medda, R., Wurm, C.A., Hein, B., Jakobs, S., Eggeling, C., and Hell, S.W. (2008) Fluorescence nanoscopy by ground-state depletion and single-molecule return. *Nat. Methods*, **5** (11), 943–945.

4
Chromophore-Assisted Light Inactivation:
A Twenty-Year Retrospective

Daniel G. Jay

4.1
Historical Perspective

Chromophore-assisted light inactivation (CALI) was developed as a means of targeted inactivation of proteins to generate a protein loss of function at the time of light excitation [1]. In its original version, light energy is targeted to destroy a protein of interest by binding it with a specific antibody labeled with the chromophore, Malachite green. Short-lived free radicals are generated when the dye is excited by laser light at 620 nm, a wavelength not well absorbed by cellular components, resulting in photochemical damage to the bound antigen with minimal collateral damage to nearest neighbors. Since its inception, CALI has been useful in ascribing *in situ* function of a large number of proteins in cells and embryos and has spawned a family of related technologies that expand the power and diversity of application of light inactivation of protein function. Based on the twentieth anniversary of its invention, it is timely to review the development of CALI technology, its use to ascribe *in situ* function, and the future of its potential application. There have been other good comprehensive reviews on this technology and the reader is directed to these particularly for methods [2–4].

The development of CALI was derived from two ideas: (i) laser ablation of whole cells *in vitro* [5] and in simple organisms such as grasshopper [6] and (ii) the promise of monoclonal antibody technology to deliver highly specific inhibitors of protein function had not been realized. Combining these two ideas led the author to consider that laser energy might be targeted to damage specific proteins via antibodies using a dye that would absorb wavelengths of light not readily absorbed by cells or tissue. The excited chromophore would release the absorbed energy rapidly and in the form of heat, thus denaturing the bound antigen. Feasibility calculations were done with Allan Oseroff (then at Wellman Labs Massachusetts General Hospital) that suggested the possibility for successful denaturation of proteins by the proposed approach. In discussions with James Foley (then at the Rowland Institute) triarylmethanes were suggested and the author selected Malachite green as the first photosensitizer for CALI, because it absorbed 620 nm

Laser Imaging and Manipulation in Cell Biology. Edited by Francesco S. Pavone
Copyright © 2010 WILEY-VCH Verlag GmbH & Co. KGaA, Weinheim
ISBN: 978-3-527-40929-7

light, which is not appreciably absorbed by cells, it relaxed mostly by vibrational modes releasing its energy as heat, and it was light stable such that the dye could undergo many cycles of excitation–relaxation. It was later determined that the mechanism of CALI was not thermal denaturation but instead is due to photo-generation of short-lived free radicals [7].

The first CALI study was applied to β-galactosidase and alkaline phosphatase using Malachite green-labeled antibodies specific for theses soluble enzymes. The 620-nm laser irradiation resulted in a targeted inactivation of enzyme function for these two proteins [1]. Increasing doses of laser light could inactivate 95% of β-galactosidase when incubated with Malachite green-labeled *anti*-β-galactosidase without signifi-cant effects to alkaline phosphatase in the same solution. Additionally in this study, the membrane associated enzyme acetylcholinesterase in erythrocyte membranes was inactivated without significant damage to distilbene disulfonate (DIDS) binding of the anion exchange protein, Band 3, as a control for damage to non-targeted proteins in the same membrane.

The first biological application of CALI was to address the *in vivo* role of a cell adhesion molecule, fasciclin I, in the stereotyped axon growth of pioneer neurons during grasshopper limb development. In this study Jay and Keshishian [8] showed that CALI of fasciclin I resulted in splitting of axons that were normally bundled as they extended along the developing limb but caused no defect in axon guidance. This study was notable in the large number of embryos analyzed.

4.2
Family of CALI-Based Technologies

Since its inception, CALI has grown into a multitude of light inactivation approaches that take advantage of diverse chromophores, localization strategies, and molecular encoded photosensitizers. Surrey *et al.* [9] established that CALI could be performed using various different chromophores beyond Malachite green, albeit with different efficacies. The commonly used fluorophore fluorescein was highly efficient in loss of β-galactosidase activity in CALI while green fluorescent protein (GFP) was about the same as Malachite green. Fluorophore-assisted light inactivation (FALI) was devel-oped [10], which took advantage of the greater efficiency of fluorescein to permit use of non-laser light sources (such as a slide projector) to simultaneously irradiate many samples in multi-well plates [11], thus providing the potential for high-throughput CALI [12, 13].

The use of GFP for CALI would have advantages because it could be expressed in cells as a genetically encoded photosensitizer and this idea was tested and exploited by Rajfur *et al.* [14]. These authors use EGFP, which has low efficiency for CALI. While not ideal, this low efficiency is potentially an advantage since low light can be used for fluorescence visualization of the targeted expressed fusion protein, followed by high intensity light exposure when and where the investigator chooses to inactivate this protein. Two other CALI technologies used the expression of a tetracysteine domain that can bind tightly to diarsenyl derivatives of fluorescein (FlAsH) and DS Red

(ReAsH) to develop FlAsH-FALI [15] and ReAsH CALI [16], respectively. FlAsH or ReAsH do not fluoresce in solution but do so when bound to a protein fusion containing the tetracysteine domain. These approaches take advantage of the higher efficiency of the FlAsH and ReAsH in CALI-mediated damage and the power of molecular genetics by expressing the encoded binding domain. Bulina *et al.* [17] isolated the novel molecularly encoded photosensitizer KillerRed, derived by the directed evolution and screening of mutations of the chromophore anm2CP. KillerRed has high efficacy and has been used for both targeted phototoxicity [18] and CALI [17]. Notably, all forms of CALI using molecularly encoded photosensitizers eliminate only the expressed fusion protein or potentially nearest neighbors such as homodimers, so strategies must be employed to eliminate endogenous protein expression by mutation [15] or interference RNA [19].

Searches for novel photosensitizers have borne fruit [17, 20] and expanded the arsenal of light inactivation probes available. For example, CALI-based approaches have used other photosensitizers such a gold nanoparticles [21] and ruthenium red [20]. In addition, novel means of targeting the damage have been developed, including small molecule binders [22, 23], RNA aptamers [24], and environmentally sensitive fluorophores [25].

4.3
Spatial Restriction of Damage

An important aspect of CALI-induced damage is its spatial restriction. The power of the technology is its ability to damage one protein exclusively when and where light is applied. The first CALI studies tested this notion with two enzymes in solution or in erythrocyte membranes but for its prudent application in living cells it was necessary to determine a half maximal radius of damage and to assess nonspecific damage in cells. Early studies using various approaches estimated half maximal radius of damage at ~15 Å. Measure of damage by moving the dyes further from the targeted protein showed marked drop off of inactivation for each intervening protein when β-galactosidase was targeted [26] and when different subunits of the T cell receptor in lymphocytes were inactivated using antibodies specific for different subunits whose nearest neighbors are known [27]. A third approach measured CALI in response to differing concentrations of free radical quenching to determine a similar value for half-maximal radius of damage [7]. A striking example of the spatial restriction of CALI *in vivo* was a control in which Malachite green-labeled secondary antibody was included with an unlabeled primary antibody specific for the axon guidance molecule ephrin A5 [28]. In contrast to direct CALI of ephrinA5, which shows guidance defects, light irradiation had no effect when the chromophore was attached to the secondary antibody and not in direct contact with the targeted protein [28]. Despite these studies that indicate that CALI is highly spatially restricted, one should use caution in interpreting results in that nonspecific damage to a nearest neighbor that is particularly sensitive to free radical damage remains a possibility. This has been illustrated in experiments using FALI.

The extent of FALI-induced damage is greater, with half maximal radius of damage of 40 Å measured by quencher analysis [10], which is consistent with the longer lifetime of singlet oxygen compared to hydroxyl radical. In assessing the spatial restriction of FALI, Horstkotte *et al.* [29] compared the quantum yield for antibody directed CALI of β-galactosidase with free fluorescein in solution and showed a two orders of magnitude increase in quantum yield for light-mediated loss of activity when fluorescein is localized near the enzyme via the antibody. However, for FALI, nonspecific collateral damage has been observed after FALI using fluorescein-labeled α-bungarotoxin (BTX) targeted by a BTX binding site engineered into different receptors [30]. Spatial restriction has been estimated at 60 Å for GFP-CALI [19] and not yet been well estimated for other forms of CALI.

4.4
Mechanism of CALI

The original conception of CALI was as a photothermal mechanism and the dye Malachite green was selected based on its efficient conversion of 620 nm light absorption into vibrational modes. The first thought that other mechanisms might be involved in CALI was a direct measure of how much of the absorbed energy was dissipated rapidly using photoacoustic calorimetry after excitation with 620 nm light of Malachite green labeled bovine serum albumin [31]. These studies showed that while the vast majority of the absorbed energy was very rapidly converted into heat, 2% was retained and could possibly be available for release via other modes [31]. Liao *et al.* [7] showed that CALI was dependent on the number of photons delivered and free radical quenchers inhibited CALI, demonstrating a photochemical mechanism for CALI. While chemical quenching experiments favored OH radical as opposed to singlet oxygen, these studies were not definitive but were consistent with the tighter radius of damage seen with CALI compared with FALI. Triplet state quenchers had no effect on CALI, suggesting that triplet state was not involved in CALI-induced free radical generation. The author examined the resultant effects on Malachite green-labeled bovine serum albumin and observed oxidized products of methionine, cysteine, and tryptophan, which are consistent with reactive oxygen species being the mediator of damage for CALI (D.G. Jay unpublished results).

More recently, Yan *et al.* [32] and Horstkotte *et al.* [29] have extended our chemical understanding of the FALI mechanism using fluorescein and FlAsH. Yan *et al.* [32] used mass spectrometry to analyze the resultant products generated after FlAsH-FALI, showing mostly modification of amino acyl side chains, especially methionine, which is easily oxidized, with some crosslinking consistent with a mechanism of local generation of reactive oxygen species. Protein crosslinking can occur and has been observed when FALI was applied to a ribosomal protein, pKi-67, which acts in ribosomal RNA synthesis [33]. Horstkotte *et al.* [29] assessed FALI of β-galactosidase using fluorescein free in solution compared to bound to *anti*-β-galactosidase. Using transient spectroscopy these authors showed the importance of the first singlet state and the generation of singlet oxygen for FALI-mediated damage and that FALI was independent of photobleaching.

Recent work has addressed the mechanism of GFP-CALI by McLean *et al.* [34], also implicating singlet oxygen and showing that the efficacy of CALI is best for EGFP among the GFP-related fluorophores tested but all these species are much less efficient than FlAsH. Recently, the crystal structure of KillerRed, the most efficient of the encoded photosensitizers, was established [35]. Its structure shows a long water-filled channel connecting the fluorophore with solvent such that molecular oxygen is accessible. This structural difference from the GFP-related molecules could explain why KillerRed is a more efficient CALI reagent than EGFP.

4.5
Micro-CALI

A particularly useful application of CALI is micro-CALI, in which proteins can be inactivated in regions within cells in the micron range using focused light through microscope optics [36]. Micro-CALI offered a novel and powerful approach to address *in situ* function in single cells and this technology was exploited to address the roles of specific proteins in neuronal growth cone in culture, including calcineurin [37], myosin V and I [38, 39], myosin II [40], talin and vinculin [41], dynactin [42], and the IP3 receptor [43–45]. These studies were able to ascertain highly localized roles for specific proteins in cell motility. For example, ezrin has a role in the leading edge of fibroblast but not in their trailing edge [46]. Thus far, micro-CALI remains the only method for loss of function with subcellular specificity. One caveat to this approach is that recovery is rapid, dependent on diffusion or trafficking of active proteins from un-irradiated regions of the cell.

4.6
Intracellular Targets of CALI

While CALI was applied to address extracellular proteins in axon guidance and neurite outgrowth [8, 28, 47], the power of CALI technology was best illustrated by a series of experiments addressing the intracellular roles of proteins in cells and tissue in a diversity of embryos and cells in culture. Much of the early application of CALI intracellularly was performed to address the function of proteins in the growth cones of primary neurons in culture. Prior to the development of CALI, experiments addressing *in situ* roles of specific proteins in primary neurons were rare. since knockdown or overexpression was not optimal in these cells, neuron-like cultured cells such as PC12 and neuroblastoma cells were the cells of choice. Defining specific functional roles for cytoskeletal proteins such as talin, vinculin [41], ERM proteins [48], myosin motors, and myosin Ib, II and 5c, [38, 40] were a major achievement of CALI-based experiments. The specificity of effects of some of these experiments was particularly interesting. For example, when contrasting two actin-associated proteins, loss of talin was found to cause fliopodial retraction while loss of vinculin caused filopodial buckling [41]. Loss of myosin V also resulted in filopodial retraction while loss of myosin Ib caused lamellipodial extension [38]. In addition, signal transduction molecules were addressed, including calcineurin [37], pp60c-src [49], and the IP3 receptor [43]. Cell adhesion molecules were also addressed Takei *et al.* [50]. Together, these CALI-based studies of protein elements from receptors to signal transduction to cytoskeletal elements provide the basis for mechanistic understanding of the action of these proteins during growth cone motility and neurite outgrowth.

The cytoskeleton has been a frequent target for CALI. In addition to the myosins, talin, vinculin, and ERM proteins previously mentioned, capping protein (CP) in fish keratocytes has been inactivated by GFP-CALI, leading to increased actin polymerization consistent with its regulator role in capping barbed ends [19]. In addition, tubulin [51], microtubules [9], and their associated proteins [52, 53] have been studied.

Many CALI experiments have resulted in changes that diminish cell behavior and there is always a concern that this diminution could have resulted from nonspecific damage despite controls. However, several CALI experiments result in increased cell activity that would be unlikely to occur from nonspecific damage. For example, loss of the tyrosine kinase pp60c-Src caused increased neurite outgrowth consistent with the application of Src kinase-specific inhibitors [49]. CALI of myosin 1b caused lamellipodial expansion in nerve growth cones [38] and repeated CALI on one side of the growth cone led to a turning of neurite outgrowth toward the laser spot [39]. When CP was replaced with EGFP-fusion to CP, light irradiation increased actin polymerization, which is consistent with the regulatory role that CP has on F-actin dynamics [19].

4.7
CALI *In Vivo*

Some of the earliest CALI-based studies were performed using grasshopper embryos *in vivo* [8, 36]. CALI was also employed to address *Drosophila* pattern formation inactivating the protein product of the pair rule gene even skipped [54]. This work showed that CALI could phenocopy a well-established loss of function mutation, showing that acute loss of this nuclear transcription factor resulted in the same embryonic defects in pattern formation as seen in a severe hypomorph. This allowed Schroder *et al.* [55] to then extrapolate this approach to address the role of even skipped in the flour beetle (*Tribolium* spp.), which lack good genetic tools and has a different mechanisms of segmental development (short germ band versus long germ band) to show that this gene's pair rule function is conserved. Other *in vivo* studies addressing the Patched protein in *Drosophila* larval eye development [56] and in embryonic chick axon guidance [28] begin to illustrate the potential for CALI-based technologies for whole organism studies.

The deployment for CALI of molecularly encoded photosensitizers that are in common use such as EGFP [14] greatly extends its *in vivo* application when spatial or temporal resolution is required to address mechanism. This is well illustrated by an elegant set of studies using FlAsH FALI addressed roles for synaptic proteins in *Drosophila* embryos. These studies took full advantage of the spatial resolution of CALI, and by rescuing null mutants with transgenes encoding proteins with FlAsH binding cassette it was possible to inactivate synaptotagmin I specifically at *Drosophila* synapses and assess the consequences of this acute loss [15]. These papers implicate synaptotagmin I in vesicle endocytosis instead of its previously considered role in exocytosis [57]. Other proteins at synapses have also been addressed such as clathrin in *Drosophila* [58].

4.8
High-Throughput Approaches

The advent of more efficient photosensitizers and robotic and automated assays led to the possibility of CALI being used in a high-throughput screen. This was realized by Eustace *et al.* [12] who used FALI in conjunction with a phage display single chain antibody library to interrogate the surface proteome of HT-1080 fibrosarcoma cells for proteins that function in invasiveness. In this function-first approach, antibodies that caused a loss of transwell invasion after FALI were used to identify their bound antigen by immunoprecipitation and mass spectrometry. This first screen identified an extracellular hsp90 alpha and subsequent studies implicated the poliovirus receptor (CD155) and neuropilin 1 in cancer invasiveness and cell migration [13, 59]. Such approaches have been applied to other cellular processes such as apoptosis [60] and are likely to have utility for discovery based screens addressing many different cellular processes. Beyond FALI-based screens, other photosensitizers may also be used and ruthenium red has been tested for its applicability in enhancing a small molecule screen [20].

4.9
Future of CALI

When CALI was developed, there were very few loss-of-function strategies available to address cellular function. Since then interference RNA strategies that decrease expression have been developed that are facile and high-throughput (reviewed in Reference [61]). CALI still provides the localized and acute loss of function not provided by RNAi approaches. CALI thus complements RNAi and has some advantages: it is not subject to genetic compensation, it is more similar to drug effects in that loss is acute inhibition and not chronic loss of expression, and it is possible to target isoforms and posttranslational modifications, for example, with phosphospecific antibodies. Subcellular micro-CALI and its application may prove valuable to address the complex networks of interactions such as those found in dendritic spines in neural connections. In addition, the temporal resolution of CALI may be useful for addressing developmental timing and for addressing genes whose loss results in lethality before the events to be studied.

The unique advantages of CALI may be combined in the future with other recent technological advances. For example, high-throughput approaches combined with molecularly encoded photosensitizers would permit the full exploitation of the fusion of molecular genetics with light-based inactivation. Also, combining micro-CALI with high-resolution molecular imaging and single molecular behavior in cells (reviewed in Reference [62]) would enhance approaches to understanding subcellular events. The array of photosensitizers now available makes multicolor-CALI possible using different chromophores and different excitations to assess the role of different proteins in a given pathway and potentially the epistasis of their roles. The use of confocal and multiphoton microscopes for CALI [63] and the use of high efficiency photosensitizers provide the possibility for the widespread use of micro-CALI throughout the biomedical research community.

References

1 Jay, D.G. (1988) Selective destruction of protein function by chromophore-assisted laser inactivation. *Proc. Natl. Acad. Sci. USA*, **85** (15), 5454–5458.

2 Eustace, B.K. and Jay, D.G. (2003) Fluorophore-assisted light inactivation for multiplex analysis of protein function in cellular processes. *Methods Enzymol.*, **360**, 649–660.

3 Hoffman-Kim, D., Diefenbach, T.J., Eustace, B.K., and Jay, D.G. (2007) Chromophore-assisted laser inactivation. *Methods Cell Biol.*, **82**, 335–354.

4 Jacobson, K., Rajfur, Z., Vitriol, E., and Hahn, K. (2008) Chromophore-assisted laser inactivation in cell biology. *Trends Cell Biol.*, **18** (9), 443–450.

5 Miller, J.P. and Selverston, A. (1979) Rapid killing of single neurons by irradiation of intracellularly injected dye. *Science*, **206**, 702–704.

6 Keshishian, H. and Bentley, D. (1983) Embryogenesis of peripheral nerve pathways in grasshopper legs. III. Development without pioneer neurons. *Dev. Biol.*, **96** (1), 116–124.

7 Liao, J.C., Roider, J., and Jay, D.G. (1994) Chromophore-assisted laser inactivation of proteins is mediated by the photogeneration of free radicals. *Proc. Natl. Acad. Sci. USA*, **91**, 2659–2663.

8 Jay, D.G. and Keshishian, H. (1990) Laser inactivation of fasciclin I disrupts axon adhesion of grasshopper pioneer neurons. *Nature*, **348** (6301), 548–550.

9 Surrey, T., Elowitz, M.B., Wolf, P.E., Yang, F., Nedelec, F., Shokat, K., and Leibler, S. (1998) Chromophore-assisted light inactivation and self-organization of microtubules and motors. *Proc. Natl. Acad. Sci. USA*, **95** (8), 4293–4298.

10 Beck, S., Sakurai, T., Eustace, B.K., Beste, G., Schier, R., Rudert, F., and Jay, D.G. (2002) Fluorophore-assisted light inactivation: a high-throughput tool for direct target validation of proteins. *Proteomics*, **2** (3), 247–255.

11 Rubenwolf, S., Niewohner, J., Meyer, E., Petit-Frere, C., Rudert, F., Hoffmann, P.R., and Ilag, L.L. (2002) Functional proteomics using chromophore-assisted laser inactivation. *Proteomics*, **2** (3), 241–246.

12 Eustace, B.K., Sakurai, T., Stewart, J.K., Yimlamai, D., Unger, C., Zehetmeier, C., Lain, B., Torella, C., Henning, S.W., Beste, G., Scroggins, B.T., Neckers, L., Ilag, L.L., and Jay, D.G. (2004) Functional proteomic screens reveal an essential extracellular role for hsp90 alpha in cancer cell invasiveness. *Nat. Cell Biol.*, **6** (6), 507–514.

13 Sloan, K.E., Eustace, B.K., Stewart, J.K., Zehetmeier, C., Torella, C., Simeone, M., Roy, J.E., Unger, C., Louis, D.N., Ilag, L.L., and Jay, D.G. (2004) CD155/PVR plays a key role in cell motility during tumor cell invasion and migration. *BMC Cancer*, **41**, 73.

14 Rajfur, Z., Roy, P., Otey, C., Romer, L., and Jacobson, K. (2002) Dissecting the link between stress fibres and focal adhesions by CALI with EGFP fusion proteins. *Nat. Cell Biol.*, **4** (4), 286–293.

15 Marek, K.W. and Davis, G.W. (2002) Transgenically encoded protein photoinactivation (FlAsH-FALI): acute inactivation of synaptotagmin I. *Neuron*, **36** (5), 805–813.

16 Tour, O., Meijer, R.M., Zacharias, D.A., Adams, S.R., and Tsien, R.Y. (2003) Genetically targeted chromophore-assisted light inactivation. *Nat. Biotechnol.*, **21** (12), 1505–1508.

17 Bulina, M.E., Chudakov, D.M., Britanova, O.V., Yanushevich, Y.G., Staroverov, D.B., Chepurnykh, T.V., Merzlyak, E.M., Shkrob, M.A., Lukyanov, S., and Lukyanov, K.A. (2006) A genetically encoded photosensitizer. *Nat. Biotechnol.*, **24** (1), 95–99.

18 Serebrovskaya, E.O., Edelweiss, E.F., Stremovskiy, O.A., Lukyanov, K.A., Chudakov, D.M., and Deyev, S.M. (2009) Targeting cancer cells by using an antireceptor antibody-photosensitizer fusion protein. *Proc. Natl. Acad. Sci. USA*, **106** (23), 9221–9225.

19 Vitriol, E.A., Uetrecht, A.C., Shen, F., Jacobson, K., and Bear, J.E. (2007) Enhanced EGFP-chromophore-assisted laser inactivation using deficient cells rescued with functional EGFP-fusion proteins. *Proc. Natl. Acad. Sci. USA*, **104** (16), 6702–6707.

20 Lee, J., Yu, P., Xiao, X., and Kodadek, T. (2008) A general system for evaluating the efficiency of chromophore-assisted light inactivation (CALI) of proteins reveals Ru (ii) tris-bipyridyl as an unusually efficient "warhead". *Mol. Biosyst.*, **4** (1), 59–65.

21 Pitsillides, C.M., Joe, E.K., Wei, X., Anderson, R.R., and Lin, C.P. (2003) Selective cell targeting with light-absorbing microparticles and

nanoparticles. *Biophys. J.*, **84** (6), 4023–4032.

22 Inoue, T., Kikuchi, K., Hirose, K., Iino, M., and Nagano, T. (2001) Small molecule-based laser inactivation of inositol 1,4,5-trisphosphate receptor. *Chem. Biol.*, **8** (1), 9–15.

23 Marks, K.M., Braun, P.D., and Nolan, G.P. (2004) A general approach for chemical labeling and rapid, spatially controlled protein inactivation. *Proc. Natl. Acad. Sci. USA*, **101** (27), 9982–9987.

24 Grate, D. and Wilson, C. (1999) Laser-mediated, site-specific inactivation of RNA transcripts. *Proc. Natl. Acad. Sci. USA*, **96** (11), 6131–6136.

25 Yogo, T., Urano, Y. *et al.* (2008) Selective photoinactivation of protein function through environment-sensitive switching of singlet oxygen generation by photosensitizer. *Proc. Natl. Acad. Sci. USA*, **105** (1), 28–32.

26 Linden, K.G., Liao, J.C., and Jay, D.G. (1992) Spatial specificity of chromophore assisted laser inactivation of protein function. *Biophys. J.*, **61** (4), 956–962.

27 Liao, J.C., Berg, L.J., and Jay, D.G. (1995) Chromophore-assisted laser inactivation of subunits of the T-cell receptor in living cells is spatially restricted. *Photochem. Photobiol.*, **62** (5), 923–929.

28 Sakurai, T., Wong, E., Drescher, U., Tanaka, H., and Jay, D.G. (2002) Ephrin-A5 restricts topographically specific arborization in the chick retinotectal projection in vivo. *Proc. Natl. Acad. Sci. USA*, **99** (16), 10795–10800.

29 Horstkotte, E., Schroder, T., Niewohner, J., Thiel, E., Jay, D.G., and Henning, S.W. (2005) Toward understanding the mechanism of chromophore-assisted laser inactivation – evidence for the primary photochemical steps. *Photochem. Photobiol.*, **81** (2), 358–366.

30 Guo, J., Chen, H., Puhl, H.L. 3rd, and Ikeda, S.R. (2006) Fluorophore-assisted light inactivation produces both targeted and collateral effects on N-type calcium channel modulation in rat sympathetic neurons. *J. Physiol.*, **576** (Pt 2), 477–479.

31 Indig, G., Jay, D.G., and Grabowski, J. (1992) The photo-induced heat-release properties of Malachite green and

Malachite green bound to protein. *Biophys. J.*, **61**, 631–638.

32 Yan, P., Xiong, Y., Chen, B., Negash, S., Squier, T.C., and Mayer, M.U. (2006) Fluorophore-assisted light inactivation of calmodulin involves singlet-oxygen mediated cross-linking and methionine oxidation. *Biochemistry*, **45** (15), 4736–4748.

33 Rahmanzadeh, R., Huttmann, G., Gerdes, J., and Scholzen, T. (2007) Chromophore-assisted light inactivation of pKi-67 leads to inhibition of ribosomal RNA synthesis. *Cell Prolif.*, **40** (3), 422–430.

34 McLean, M.A., Rajfur, Z., Chen, Z., Humphrey, D., Yang, B., Sligar, S.G., and Jacobson, K. (2009) Mechanism of chromophore assisted laser inactivation employing fluorescent proteins. *Anal. Chem.*, **81** (5), 1755–1761.

35 Carpentier, P., Violot, S., Blanchoin, L., and Bourgeois, D. (2009) Structural basis for the phototoxicity of the fluorescent protein KillerRed. *J. Am. Chem. Soc.*, **131** (50), 18063–18065.

36 Diamond, P., Mallavarapu, A., Schnipper, J., Booth, J., Park, L., O'Connor, T.P., and Jay, D.G. (1993) Fasciclin I and II have distinct roles in the development of grasshopper pioneer neurons. *Neuron*, **11** (3), 409–421.

37 Chang, H.Y., Takei, K., Sydor, A.M., Born, T., Rusnak, F., and Jay, D.G. (1995) Asymmetric retraction of growth cone filopodia following focal inactivation of calcineurin. *Nature*, **376** (6542), 686–690.

38 Wang, F.S., Wolenski, J.S., Cheney, R.E., Mooseker, M.S., and Jay, D.G. (1996) Function of myosin-V in filopodial extension of neuronal growth cones. *Science*, **273** (5275), 660–663.

39 Wang, F.S., Liu, C.W., Diefenbach, T.J., and Jay, D.G. (2003) Modeling the role of myosin 1c in neuronal growth cone turning. *Biophys. J.*, **85** (5), 3319–3328.

40 Diefenbach, T.J., Latham, V.M., Yimlamai, D., Liu, C.A., Herman, I.M., and Jay, D.G. (2002) Myosin 1c and myosin IIB serve opposing roles in lamellipodial dynamics of the neuronal growth cone. *J. Cell Biol.*, **158** (7), 1207–1217.

41 Sydor, A.M., Su, A.L., Wang, F.S., Xu, A., and Jay, D.G. (1996) Talin and vinculin play distinct roles in filopodial motility in the neuronal growth cone. *J. Cell Biol.*, **134** (5), 1197–1207.

42 Abe, T.K., Honda, T., Takei, K., Mikoshiba, K., Hoffman-Kim, D., Jay, D.G., and Kuwano, R. (2008) Dynactin is essential for growth cone advance. *Biochem. Biophys. Res. Commun.*, **372**, 418–422.

43 Takei, K., Shin, R.M., Inoue, T., Kato, K., and Mikoshiba, K. (1998) Regulation of nerve growth mediated by inositol 1,4,5-trisphosphate receptors in growth cones. *Science*, **282** (5394), 1705–1708.

44 Yogo, T., Kikuchi, K., Inoue, T., Hirose, K., Iino, M., and Nagano, T. (2004) Modification of intracellular Ca2 + dynamics by laser inactivation of inositol 1,4,5-trisphosphate receptor using membrane-permeant probes. *Chem. Biol.*, **11** (8), 1053–1058.

45 Iketani, M., Imaizumi, C., Nakamura, F., Jeromin, A., Mikoshiba, K., Goshima, Y., and Takei, K. (2009) Regulation of neurite outgrowth mediated by neuronal calcium sensor-1 and inositol 1,4,5-trisphosphate receptor in nerve growth cones. *J. Neurosci.*, **161** (3), 743–752.

46 Lamb, R.F., Ozanne, B.W., Roy, C., McGarry, L., Stipp, C., Mangeat, P., and Jay, D.G. (1997) Essential functions of ezrin in maintenance of cell shape and lamellipodial extension in normal and transformed fibroblasts. *Curr. Biol.*, **7** (9), 682–688.

47 Muller, B.K., Jay, D.G., and Bonhoeffer, F. (1996) Chromophore-assisted laser inactivation of a repulsive axonal guidance molecule. *Curr. Biol.*, **6** (11), 1497–1502.

48 Castelo, L. and Jay, D.G. (1999) Radixin is involved in lamellipodial stability during nerve growth cone motility. *Mol. Biol. Cell*, **10** (5), 1511–1520.

49 Hoffman-Kim, D., Kerner, J.A., Chen, A., Xu, A., Wang, T.F., and Jay, D.G. (2002) pp60(c-src) is a negative regulator of laminin-1-mediated neurite outgrowth in chick sensory neurons. *Mol. Cell Neurosci.*, **21** (1), 81–93.

50 Takei, K., Chan, T.A., Wang, F.S., Deng, H., Rutishauser, U., and Jay, D.G. (1999)

The neural cell adhesion molecules L1 and NCAM-180 act in different steps of neurite outgrowth. *J. Neurosci.*, **9** (21), 9469–9479.

51 Keppler, A. and Ellenberg, J. (2009) Chromophore-assisted laser inactivation of alpha- and gamma-Tubulin SNAP-tag fusion proteins inside living cells. *ACS Chem. Biol.*, **4** (2), 127–138.

52 Liu, C.W., Lee, G., and Jay, D.G. (1999) Tau is required for neurite outgrowth and growth cone motility of chick sensory neurons. *Cell Motil. Cytoskeleton*, **43** (3), 232–242.

53 Mack, T.G., Koester, M.P., and Pollerberg, G.E. (2000) The microtubule-associated protein MAP1B is involved in local stabilization of turning growth cones. *Mol. Cell Neurosci.*, **15** (1), 51–65.

54 Schroeder, R., Tautz, D., and Jay, D.G. (1996) Chromophore-assisted laser inactivation of even skipped in Drosophila phenocopies genetic loss of function. *Dev. Genes Evol.*, **206**, 86–88.

55 Schroder, R., Jay, D.G., and Tautz, D. (1999) Elimination of EVE protein by CALI in the short germ band insect Tribolium suggests a conserved pair-rule function for even skipped. *Mech. Dev.*, **80** (2), 191–195.

56 Schmucker, D., Su, A.L., Beermann, A., Jackle, H., and Jay, D.G. (1994) Chromophore-assisted laser inactivation of patched protein switches cell fate in the larval visual system of Drosophila. *Proc. Natl. Acad. Sci. USA*, **91** (7), 2664–2668.

57 Poskanzer, K.E., Marek, K.W., Sweeney, S.T., and Davis, G.W. (2003) Synaptotagmin I is necessary for compensatory synaptic vesicle endocytosis in vivo. *Nature*, **426**, 559–563.

58 Heerssen, H., Fetter, R.D., and Davis, G.W. (2008) Clathrin dependence of synaptic-vesicle formation at the Drosophila neuromuscular junction. *Curr. Biol.*, **18** (6), 401–409.

59 Bagci, T., Wu, J.K., Pfannl, R., Ilag, L.L., and Jay, D.G. (2009) Autrocrine semaphorin 3A signaling promotes glioblastoma dispersal. *Oncogene*, **28** (40), 3537–3550.

60 Hauptschein, R.S., Sloan, K.E., Torella, C., Moezzifard, R., Giel-Moloney, M., Zehetmeier, C., Unger, C., Ilag, L.L., and Jay, D.G. (2005) Functional proteomic screen identifies a modulating role for CD44 in death receptor-mediated apoptosis. *Cancer Res.*, **65** (5), 1887–1896.

61 Boutros, M. and Ahringer, J. (2008) The art and design of genetic screens: RNA interference. *J. Nat. Rev. Genet.*, **9** (7), 554–566.

62 Rosivatz, E. (2008) Imaging the boundaries-innovative tools for microscopy of living cells and real-time imaging. *J. Chem. Biol.*, **1** (1–4), 3–15.

63 Tanabe, T., Oyamada, M., Fujita, K., Dai, P., Tanaka, H., and Takamatsu, T. (2005) Multiphoton excitation-evoked chromophore-assisted laser inactivation using green fluorescent protein. *Nat. Methods*, **2** (7), 503–505.

5
Photoswitches

Andrew A. Beharry and G. Andrew Woolley

5.1
Introduction

Laser manipulation of specific molecular events can be a powerful tool in cell biology. Light is a tool that is easy to control and is relatively non-invasive. It provides high spatial and temporal resolution allowing one to investigate rapid biochemical processes confined to small regions of the cell. While some processes of interest may be naturally light sensitive, in most cases light-sensitive activity must be engineered into the system. Introduction of light sensitivity may be accomplished by one of two general approaches. The first is a chemical approach, involving the synthesis and application of small molecule chromophores that undergo a photo-chemical process leading to alteration in biochemical activity. The second approach involves engineering natural light-sensitive proteins so that they perform a desired molecular task.

The first synthetic small molecule chromophores to find wide application in biochemical systems were "caged" compounds, beginning with the introduction of a "caged" ATP [1, 2]. Caged ATP is biologically inert ("caged") in the dark. Absorption of a photon triggers a photolysis reaction releasing active ATP, which has numerous biological effects depending on the system in question. Numerous caged molecules have been developed since this study, including caged ions, neurotransmitters, DNA, RNA, and proteins [3, 4]. Caged neurotransmitters (e.g., caged glutamate) have been particularly useful in a wide range of studies in neurobiology [5, 6]. Numerous reviews have dealt with the synthesis and application of caged compounds [3, 4, 7, 8], including the excellent volume *Dynamic Studies in Biology* by Goeldner and Givens [9].

Caged compounds undergo essentially *irreversible* photochemistry, leading to their designation as "phototriggers." That is, once photolysis (or phototriggering) has occurred, the biomolecule will always be active, unless it is degraded or metabolized in some manner. The term "photoswitch," however, refers to molecules that undergo inherently *reversible* photochemistry such that the active and inactive states of the biomolecule can be produced in multiple rounds. Examples include light induced isomerization processes (cis/trans) or light induced ring-opening/ring-closing

Laser Imaging and Manipulation in Cell Biology. Edited by Francesco S. Pavone
Copyright © 2010 WILEY-VCH Verlag GmbH & Co. KGaA, Weinheim
ISBN: 978-3-527-40929-7

processes. In principle, reversible photochemistry can be used to control cellular activity as a function of time in more complex ways than can be accomplished with irreversible photochemistry. For example, induction of gene expression could be turned on and off at specific times and locations in cells or tissues with a light switchable transcription factor, whereas caged DNA would permit light-triggered turning-on, or turning-off, of gene expression only. This short chapter will focus on recent developments of both synthetic and natural photoswitches intended for use *in vivo*.

5.2
Synthetic Photoswitches

Synthetic photoswitches fall into several groups based on the type of photochemistry they undergo. These compounds have been explored by chemists and physicists interested in applications ranging from thin film manipulation to optical data storage (e.g., [10, 11]). Their application to biology in fact predates the introduction of caged compounds [12]. Recently their application to complex biological systems has seen a resurgence. The properties of the main classes of photoswitches have been reviewed extensively [13, 14]. In general, to be suitable tools for manipulating cells, photoswitches must be stable in the cellular environment and undergo photochemistry at wavelengths compatible with cells and tissues (>350 nm) at moderate light intensities. Several photoswitches undergo photochemistry at short ultraviolet wavelengths, a feature that is less critical for those interested in optical memory storage for example. Application of the photoswitch is most straightforward if there are only two interconvertible states (e.g., cis and trans isomers). Certain photosensitive molecules exhibit more than two states (e.g., ring-closed/ring-open-cis/ring-open-trans). In such cases, the biological effect may be more difficult to design and/or control. In addition, some photoswitches undergo rapid thermal reversion, making them useless for optical data storage; such thermal reversibility may be useful, however, for certain biological applications [15, 16]. Another consideration is the completeness of photoswitching. Production of 100% of at least one of the isomers is ideal; otherwise the system will always be partially on or off. This, in turn, will limit the fold change in activity of the biomolecule [17]. Finally, the ease of synthesis and subsequent conjugation of the photoswitch to biomolecules should be considered. For example, certain classes of switches such as the fulgides and hemithioindigos are synthetically challenging [18–20], while synthesis of a wide range of azobenzene derivatives has been reported [14].

The application of synthetic photoswitches to biological targets has so far focused mostly on achieving photocontrol *in vitro*. Examples include photocontrolled DNA binding proteins [21–23], anti-apoptotic proteins [24], antibody/antigen interactions [25], and DNA/RNA aptamers [26, 27]. Our own studies have focused primarily on the photocontrol of elements of protein structure such as α-helices and β-sheets [28, 29]. It is hoped that, by developing general approaches to photocontrol of protein structure, photocontrol of function will be achievable for a wide variety of targets. The most challenging aspect of this endeavor is to understand how to effectively couple the photochemical changes the chromophore undergoes, to a change in structure and ultimately function of a targeted biomolecule. In many cases only moderate changes in activity have been reported upon isomerization. Natural photoswitchable systems, however, can show large changes in activity, indicating that it is certainly possible to couple changes in the chromophore to a protein functional change. Achievements and recent progress in the application of photoswitches to biological targets *in vitro* have been summarized in excellent reviews [30–32]. Given the pace of these developments, it is anticipated that a range of *in vivo* applications of such designed photoswitchable biomolecules will be seen soon.

Applying these chemically modified biomolecules *in vivo* requires the photoswitch be delivered to appropriate sites. This may involve microinjection of the photo-switchable biomolecule into single cells [3], the use of cell uptake peptides [33], electroporation [34], bead loading [35], or even the genetically encoded incorporation of photosensitive amino acids into proteins via modification of the biosynthetic machinery [36, 37]. Thus far, there are only a few studies that have applied a synthetic photoswitch to a target biomolecule and photocontrolled its function in a living system [38–41]. Each of these studies has employed the azobenzene chromophore as the photoswitch, perhaps because it is the simplest molecule that exhibits, to some degree, the characteristics required for a biological photoswitch listed above.

The azobenzene molecule exists in trans and cis conformations (Figure 5.1). The trans state is more stable in the dark by approximately 10 kcal mol^{-1} [42]. The trans and cis isomers have distinct but overlapping absorption spectra [43]. Irradiation of the trans isomer at 337 nm produces an excited state that relaxes to either the trans or cis isomer. Since the cis isomer absorbs weakly at 337 nm, continuous irradiation at 337 nm produces a photostationary state that is approximately 85% cis. Irradiation of the cis isomer at wavelengths longer than 450 nm, where the trans form absorbs only weakly, produces greater than 95% trans. Additionally, cis-azobenzene will undergo thermal relaxation to the trans form in the dark with a half-life of approximately two days at room temperature [43]. Modifications of azobenzene (e.g., addition of electron-donating substituents on the rings) can alter both the wavelengths for photoisomerization as well as the thermal relaxation rate. For example, near- UV (370 nm), blue (450 nm), cyan (480 nm), and green (530 nm) absorbing switches have been reported [44]. In addition, thermal relaxation rates can also be tuned from days to milliseconds at room temperature [15, 44]. Importantly, a change in shape and distance occurs upon azobenzene photoisomerization since the cis (bent) form has a

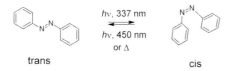

trans   cis

Figure 5.1 Azobenzene photoisomerization. The trans isomer of unsubstituted azobenzene absorbs maximally at 337 nm and irradiation at this wavelength can cause photoisomerization to the cis isomer. Thermal relaxation or irradiation at 450 nm leads to recovery of the trans isomer.

shorter end-to-end distance than the trans (extended) form. Furthermore, the cis form has a dipole moment of approximately 3 debye whereas the trans form has a dipole moment near zero [45]. Thus, due to the ease of synthesis, the production of a substantial amount of each isomer and the difference in properties between cis and trans, azobenzene has been most commonly used to alter the activity of biomolecules.

Initial investigations exploited the shape change of azobenzene to design reversible inhibitors of enzymes [12, 46, 47] and ion channels [48]. For example, a free photoswitchable *trans*-3,3'-bis[α-(trimethylammonium)methyl]azobenzene dibromide (Bis-Q) ligand (Figure 5.2) activates the nicotinic acetylcholine receptor (nAChR) at concentrations less than 0.1 μM. Photoisomerization to cis-Bis-Q by 330 nm light exposure resulted in a potency less than 1% that of trans-Bis-Q [49]. More recently, Trauner and colleagues reported a "reversibly caged" glutamate in which an azobenzene derivative was linked to the glutamate neurotransmitter [50]. Whole-cell current clamp recording of cultured rat hippocampal neurons in the presence of the reversibly caged glutamate permitted photocontrol of neuronal firing patterns in the ionotropic glutamate receptor (iGluR). In both examples, the isomerization alters the binding affinity of the ligand for the receptor [50].

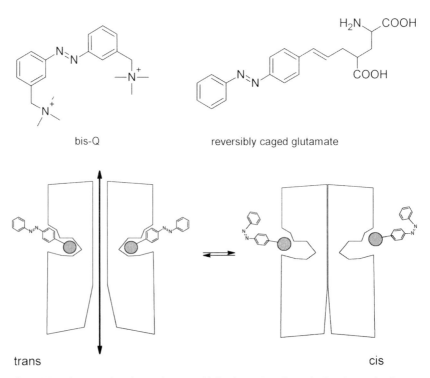

bis-Q   reversibly caged glutamate

trans   cis

Figure 5.2 Photoswitches that act by reversibly binding to ion channels. One isomer binds more tightly than the other in each case, acting as an agonist and opening the channel (indicated by the double-headed arrow showing ion flux).

Free photoswitchable ligands are rapidly diffusible and will target all receptors in a given biological preparation. It may be desirable to specifically target designated cell types or subcellular locations. In such cases, covalent modification of target proteins is preferred. Several studies of azobenzene modified ion channels have been reported, in which azobenzene isomerization is used to directly control the open/closed state of the channel. The first light-sensitive channel was developed in the absence of protein structural data. With only the knowledge that a cysteine residue was located near the acetylcholine binding site of nAChR, Erlanger and colleagues tethered 3-(α-bromo-methyl)-3'-[α-(trimethylammonium)methyl]azobenzene bromide (QBr) (Figure 5.3) after first reducing the membranes with dithiothreitol. The cis isomer of QBr exhibited lower levels of activation, an effect ascribed to the shortened length of the cis isomer so that the quaternary ammonium ion was unable to bind to its receptor [51]. This photoswitchable system provided extensive information on the mechanistic details of nAChR activation [38]. The applicability to other receptors was limited due to the requirement for a native cysteine near the binding pocket.

Later, with the advent of genetic manipulation, this tethering approach was substantially extended and generalized. Since selective membrane permeability to specific ions underlies the electrical excitability of cells, control of permeability by light permits control of excitability – a powerful tool in neuroscience to dissect the role of specific cells in neuronal circuits. To create generally applicable photochemical

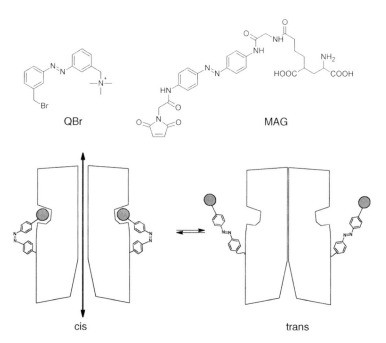

Figure 5.3 Photoswitchable ligands covalently tethered to specific ion channels. One isomer of MAG permits ion flux (indicated by the double-headed arrow) whereas the other inhibits it (either directly or allosterically). Whether the cis or trans form is more active depends on the specific Photoswitchable ligand.

tools for controlling cellular excitability Kramer, Trauner, Isacoff, and colleagues used structure-based design to develop a light-gated K^+ channel. Using the available crystal structure from a related K^+ channel and the knowledge that the channel could be completely blocked with a tetraethylammonium (TEA) compound, a thiol-reactive maleimide azobenzene derivative containing a quaternary ammonium blocker was placed at a specific distance away from the pore [52]. Long wavelength light (500 nm) converts azobenzene into the extended trans conformation, allowing the blocker to reach the pore, while short-wavelength light (390 nm) produces the shorter cis isomer, thereby removing the blocker and allowing conduction. Their light-gated construct, termed SPARK (synthetic photoisomerizable azobenzene-regulated K^+ channel), allows for photoinhibition of action potentials (hyperpolarization) as demonstrated by exogenous expression of the channel in hippocampal neurons. Thus, the photoswitchable tethered ligand (PTL) along with genetic engineering, allows for selective introduction of photoswitchable ion channels in target cells [52]. Variations in SPARK were made by mutation within the pore, converting the K^+ selective channel into a nonselective cation channel (permeable to Na^+ and Ca^+) [53]. In this case, exposure to 390 nm light (unblocking) resulted in rapid depolarization of the membrane, triggering action potentials in neurons. Together, both forms of SPARK (hyper- and depolarizing) can be used to induce opposite effects in neurons when exposed to the same wavelength of light [53].

A tethered version of the light-gated iGluR (LiGluR) was developed by the same authors [54]. In this construct, maleimide azobenzene containing glutamate (termed, MAG; right-hand structure in Figure 5.3) was first tested systematically at different attachment locations around the glutamate binding pocket. In the most effective construct, the cis conformation of azobenzene positions the glutamate ligand closer to the binding pocket, thereby increasing the effective local ligand concentration relative to the trans form (Figure 5.3). Since the "on" state is the cis form, a brief pulse of 380 nm light keeps the receptors open in the dark for minutes due to the slow thermal relaxation rate from cis to trans. The "memory" of the channel reduces illumination time thereby limiting photodamage to the cell [54]. The approach of tethering a photoswitchable ligand to an allosteric protein is likely a general and effective way of conferring photoswitchability. This concept has been reviewed recently by Gorostiza and Isacoff [39].

In some types of cells or tissues exogenous gene expression may not be possible. To overcome this problem, Kramer and colleagues have developed a photoswitchable affinity label (PAL) [55]. In this approach, similar in concept to activity based protein profiling [56], the high affinity of ligand (the azobenzene quaternary ammonium compound) for the pore of K^+ channels brings an electrophilic reactive group close to endogenous nucleophilic residues. PAL was found to retain the ability to activate and block ion flow in the cis and trans form, respectively, by covalently tethering to endogenous channels, thereby avoiding any genetic manipulation [55].

These approaches to the development of photoswitchable ion channels may be applicable to the development of other photoswitchable proteins. To date only extracellular targets have been successfully modified *in vivo* [39–41]. Intracellular targets require the photoswitch to be membrane permeable as well as stable inside cells. Although certain azobenzene derivatives appear sensitive to reduction by intracellular thiols [57], others do not [44]. An encouraging indication that intracellular targets may be made photoswitchable is the report of Bose *et al.* [37]. In this study, an orthogonal tRNA-aminoacyl tRNA synthetase pair was evolved to selectively incorporate azobenzene as an amino acid side chain (i.e., *p*-phenylazo-phenylalanine). Using *in vivo* nonsense codon suppression, the azobenzene amino acid was introduced into the *Escherichia coli* catabolite activator protein, CAP [37]. The photoswitchable-CAP bound DNA with high affinity in the trans form and weakly in the cis form when tested *in vitro*. Although the photoswitchable protein was characterized *in vitro*, the study did illustrate selective conjugation and stability of an azobenzene photoswitch *in vivo*. Extension of this approach to mammalian cells is likely possible [58].

5.3
Natural Photoswitches

A second strategy for photocontrolling function *in vivo* involves co-opting a natural light sensitive protein to control a desired chemical process [59, 60]. Perhaps the best-known natural light-sensitive protein is rhodopsin, a protein found in the retina [61]. Bound deep within the rhodopsin pocket is a covalently-linked small molecule chromophore called retinal. Absorption of a photon isomerizes the initially 11-*cis*-retinal to the all-trans conformation. The wavelength of light that triggers photo-isomerization is tuned by the protein microenvironment surrounding the retinal. Photoisomerization of the chromophore subsequently causes a conformational change in the opsin protein that sets off a signal cascade that ultimately affects ion channels in the rod cell membrane, thereby conferring light-sensitivity in vision.

Opsin-like proteins are found in a range of organisms [62]. Channelrhodopsin-1 (ChR1), found in the green alga *Chlamydomonas reinhardtii*, exhibits light-gated conductance that is selectively permeable for protons [63]. Channelrhodopsin-2 (ChR2), isolated from the same species, has characteristics of a light-gated cation-selective (e.g., Na^+ and Ca^{2+}) channel (Figure 5.4) [64]. Finally, halorhodopsin from *Natronomonas pharaonis* (NpHR) acts as a light-driven Cl^- pump [65]. The functions of the latter two proteins have been of particular interest in the field of neuroscience since; as noted with azobenzene modified channels described earlier, control of ion permeability by light can permit control of cellular excitability [66].

The usefulness of ChR2 and NpHR for optically probing and manipulating neural function has been described in an excellent review by Deisseroth and colleagues [67]. Both ChR2 and NpHR can be expressed heterologously in mammalian cells. Initially, it was expected that functional channels would require addition of the co-factor retinal or introduction of biosynthetic pathways to permit its synthesis. Surprisingly (but fortunately), this turned out not to be the case, at least for several interesting cell types, including mammalian neurons. Expression of ChR2 in mammalian cells causes rapid depolarization of the cell membrane after light exposure at 460 nm. The opposite effect occurs with NpHR, such that light exposure at 580 nm causes rapid

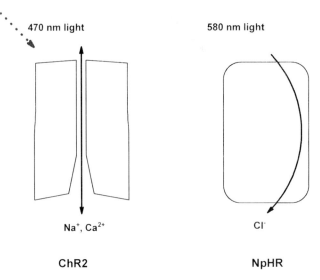

470 nm light   580 nm light

Na⁺, Ca²⁺   Cl⁻

ChR2   NpHR

Figure 5.4 Schematic of channelrhodopsin-2 (ChR2) and the halorhodopsin (NpHR) pump. Following illumination with blue light (activation maximum approx 470 nm), ChR2 allows the entry of cations (mostly Na^+ and very low levels of Ca^{2+}) into the cell, triggering action potential spiking. NpHR is activated by yellow light illumination (activation maximum approx 580 nm), allowing the entry of Cl^- anions that cause hyperpolarization and inhibit action potential firing.

hyperpolarization [67]. Since the activation wavelengths for ChR2 and NpHR are greater than 100 nm apart, when co-expressed in neurons, NpHR functions as the "yin to ChR2's yang" for silencing neuronal activity [65].

Using ChR2, Han *et al.* have developed a protocol to investigate the role of specific neuron cell types in a nonhuman primate brain. ChR2 was expressed in frontal cortex excitatory neurons of monkeys. Target cell-specificity was achieved by using a lentivirus carrying the gene for ChR2 and α-CaMKII promoter. Activation of the excitatory neurons with millisecond precision led to activation of certain neurons and suppression of activity in other types of neurons [68]. Using an adeno-associated virus as the vector, Pan and colleagues delivered the ChR2 gene nonselectively to inner retinal neurons in photoreceptor-deficient mice [69]. ChR2 expression restored the retina's ability to encode and transmit light signals to the visual cortex. Roska and colleagues have extended this work by specifically targeting ChR2 to retinal ON bipolar cells by using a promoter encoding for the ON-bipolar-cell specific glutamate receptor [70]. The engineered ON bipolar cells induced light-dependent spiking activity in ganglion cells and enhanced visual behavioral tasks in mice containing degenerated retinas. Although several challenges still exist, these studies suggest that ChR2-based stimulation techniques have the potential to restore vision in humans having retinal degeneration.

Several improvements have been made to the natural photoswitches ChR2 and NpHR since their initial discovery. By improving the signal peptide sequence and adding an ER export signal to NpHR-YFP, Zhao *et al.* dramatically increased the membrane expression of NpHR-YFP [71]. Gradinaru *et al.* have identified a combination of two motifs, an N-terminal signal peptide and a C-terminal ER export sequence, that markedly promoted membrane localization and ER export defined of NpHR. The new enhanced NpHR (eNpHR) allows safe, high-level expression in mammalian neurons, without toxicity and with augmented inhibitory function, *in vitro* and *in vivo* [72]. ChR2 shows inactivation during persistent light stimulation, which complicates optical driving of electrical responses in neurons [73]. By making chimeras of ChR1 and ChR2, combined with site-directed mutagenesis, Lin *et al.* have developed a ChR variant, named ChEF, that exhibits significantly less inactivation during persistent light stimulation. Point mutation of Ile(170) of ChEF to Val (yielding "ChIEF") accelerated the rate of channel closure while retaining reduced inactivation, leading to more consistent responses when the photoswitch was stimulated above 25 Hz. ChEF and ChIEF allow more precise temporal control of depolarization, and can induce action potential trains that more closely resemble natural spiking patterns [73]. Finally, a redshifted version of ChR2 was identified (VChR1) from *Volvox carteri* (VChR1) that can drive neuronal spiking at 589 nm, thereby expanding the range of switching wavelengths used for generation of action potentials [74].

Recently, a method has been developed by Deisseroth and colleagues that significantly extends the natural photoswitch strategy for controlling cellular excitability described above [75]. These authors realized that since rhodopsin is a member of the large family of G-protein-coupled-receptors (GPCRs), and since other GPCRs also play important roles in cellular behavior, chimeras of rhodopsin and other GPCRs (termed, optoXRs) may permit optical control of various cellular functions. These optoXRs consist of an extracellular/transmembrane rhodopsin domain and a specific type of intracellular GPCR domain (e.g., β2 receptor or α1a receptor). Depending on the optoXR expressed in cells, light-dependent increases in specific signaling molecules (e.g., cAMP for opto-β2AR- and IP3 for opto-α1AR-expressing cells) was observed at levels comparable to activation of the corresponding native GPCR [75]. Thus, this new technique provides optical control of cellular responses that are mediated by factors other than firing activity in neurons.

Another sensory photoreceptor whose function has been exploited to probe biochemical processes is the class of BLUF-proteins (blue light photoreceptors using FAD). Watanabe and colleagues have discovered a photoactivated adenylyl cyclase (PAC) in the organism *Euglena gracilis* [76]. PAC can be successfully expressed in multiple systems such as *Xenopus laevis* oocytes and *Drosophila melanogaster* [77]. Blue light (~480 nm) exposure enhanced adenylyl cyclase activity, allowing for optical control of intracellular levels of cAMP. No addition of the FAD chromophore is needed and sufficient amounts of the PAC substrate ATP are readily available in the cell. Since cAMP is required for many proteins (e.g., PKA, CNG channels), this tool holds promise for mapping signaling pathways. A current setback, however, is the high background (dark) activity. Future genetic manipulation may reveal mutants with suppressed resting activity [77].

Phytochromes are a family of plant photoreceptors [78]. Light sensitivity is conferred by a covalently-linked tetrapyrrole chromophore (e.g., phytochromobilin).

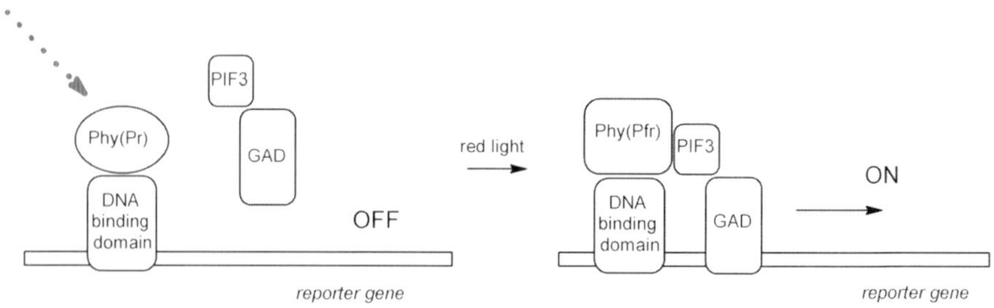

Figure 5.5 The two states of a photoswitchable promoter system developed by Shimizu-Sato *et al.* [79]. The system consists of a DNA-binding domain fused to a phytochrome chromophore domain (Phy) and PIF3 basic helix–loop–helix protein fused to a transcriptional activation domain. The Phy domain's conformation changes upon absorption of a red photon from inactive Pr to active Pfr. Pfr then interacts with PIF3, bringing the associated transcriptional activation domain into position to activate transcription from a downstream reporter [80].

They can interconvert between red-absorbing (designated Pr) and far-red-absorbing (designated Pfr) forms depending on the wavelength of light they are exposed to [78]. In addition, the active Pfr conformer can interact with specific partner proteins. For example, absorption of a red photon converts the inactive Pr (dark) form of phytochrome B (PhyB) into its active Pfr conformation. PhyB(Pfr) can then complex with the transcription factor PIF3; this interaction can be reversed upon absorption of a far-red photon.

This light-dependent interaction was first exploited by Quail and colleagues for the construction of a photoswitchable gene promoter system [79]. PhyB was fused to a GAL4 DNA-binding domain (GBD) and PIF3 was fused to a GAL4-activating domain (GAD) (Figure 5.5). When co-expressed in yeast in the presence of exogenous chromophore, red-light absorption drove expression of a reporter gene, while far-red light turned it off. Gene expression levels could be regulated by precise control of the light intensity, thereby demonstrating a dose-dependent response [79].

Muir and Tyszkiewicz [81] have integrated the photoswitchable PhyB-PIF3 interaction into a conditional protein splicing (CPS) system they developed previously [82]. Red light illumination led to rapid protein splicing, producing a new protein product directly in yeast cells. Later, Rosen and colleagues used PhyB-PIF3 to target a GTPase-effector pair [83]. A fusion of PhyB with Cdc42(GDP) and a fusion of PIF3 with WASP were expressed in bacteria together with the enzymes heme oxygenase and bilin reductase. These latter two enzymes biosynthesize the bilin chromophore, starting from heme. Upon red light exposure, the PhyB–PIF3 interaction drove effective interaction of Cdc42(GDP) and WASP, leading to WASP activation and actin assembly *in vitro*. Far-red light reversed this effect by disrupting binding and thereby WASP activation. The authors suggest this approach could be extended to target a range of GTPase-effector pairs and could provide a powerful approach for probing cell signaling [83].

Moffat and colleagues have reported recently some imaginative studies in which the blue light sensitive LOV domain was linked to other proteins to confer light sensitivity on them. In one study, a LOV domain was fused to a DNA-binding domain through an α-helical linker. Illumination with blue light resulted in enhanced DNA-binding, suggesting the α-helix can serve as an "allosteric lever arm." Their rational design strategy suggests a general scheme for coupling a LOV domain to a range of proteins [84]. In another study, the normal oxygen-sensitive domain of the histidine kinase FixL was replaced by a LOV photosensor domain. The kinase fusion had reduced phosphorylation activity (∼1000-fold) when exposed to light *in vitro*. Expression in *E. coli* allowed for photocontrol of gene expression [85].

The development of natural photoswitches as tools seems likely to grow quickly as new light sensitive modules are discovered via genome sequencing efforts. Directly co-opting a natural photoswitch for use in a particular case may be possible through imaginative use of chimeras but inevitably will require minor or perhaps major protein engineering efforts. The power of protein engineering to modify (either through structure based design or directed evolution) the properties of light sensitive proteins is considerable, as evidenced by successful efforts with fluorescent proteins [86].

One advantage of the synthetic photoswitch approach is that the properties of the chromophore can often be predicted based on empirical rules or direct calculations. If the structure of the target protein is known then photocontrol of that structure may become possible in a more direct manner via a synthetic approach. Finally, as Kramer and colleagues point out, there are numerous cases where genetic manipulation is simply not possible so that adding small molecules targeted to specific sites, as with the photoswitchable affinity label (PAL) approach, may be the most effective alternative.

References

1 Kaplan, J.H., Forbush, B. 3rd, and Hoffman, J.F. (1978) Rapid photolytic release of adenosine 5′-triphosphate from a protected analogue: utilization by the Na:K pump of human red blood cell ghosts. *Biochemistry*, **17**, 1929–1935.

2 McCray, J.A., Herbette, L., Kihara, T., and Trentham, D.R. (1980) A new approach to time-resolved studies of ATP-requiring biological systems; laser flash photolysis of caged ATP. *Proc. Natl. Acad. Sci. USA*, **77**, 7237–7241.

3 Lee, H.M., Larson, D.R., and Lawrence, D.S. (2009) Illuminating the chemistry of life: design, synthesis, and applications of "caged" and related photoresponsive compounds. *ACS Chem. Biol.*, **4**, 409–427.

4 Ellis-Davies, G.C. (2007) Caged compounds: photorelease technology for control of cellular chemistry and physiology. *Nat. Methods*, **4**, 619–628.

5 Fedoryak, O.D., Sul, J.Y., Haydon, P.G., and Ellis-Davies, G.C. (2005) Synthesis of a caged glutamate for efficient one- and two-photon photorelease on living cells. *Chem. Commun.*, 3664–3666.

6 Matsuzaki, M., Honkura, N., Ellis-Davies, G.C., and Kasai, H. (2004) Structural basis of long-term potentiation in single dendritic spines. *Nature*, **429**, 761–766.

7 Lawrence, D.S. (2005) The preparation and in vivo applications of caged peptides and proteins. *Curr. Opin. Chem. Biol.*, **9**, 570–575.

8 Marriott, G., Roy, P., and Jacobson, K. (2003) Preparation and light-directed activation of caged proteins. *Methods Enzymol.*, **360**, 274–288.

9 Goeldner, M. and Givens, R. (2005) *Dynamic Studies in Biology*, Wiley-VCH Verlag GmbH, Weinheim, p. 557.

10 Yamada, M., Kondo, M., Mamiya, J., Yu, Y., Kinoshita, M., Barrett, C.J., and Ikeda, T. (2008) Photomobile polymer materials: towards light-driven plastic motors. *Angew Chem. Int. Ed. Engl.*, **47**, 4986–4988.

11 Matharu, A.S., Jeeva, S., and Ramanujam, P.S. (2007) Liquid crystals for holographic optical data storage. *Chem. Soc. Rev.*, **36**, 1868–1880.

12 Kaufman, H., Vratsanos, S.M., and Erlanger, B.F. (1968) Photoregulation of an enzymic process by means of a light-sensitive ligand. *Science*, **162**, 1487–1489.

13 Willner, I. and Willner, B. (1993) Chemistry of photobiological switches, in *Biological Applications of Photochemical Switches* (ed. H. Morrison), John Wiley & Sons, Inc., New York, pp. 1–110.

14 Durr, H. and Bouas-Laurent, H. (2003) *Photochromism: Molecules and Systems*, Elsevier, Amsterdam, p. 1044.

15 Beharry, A.A., Sadovski, O., and Woolley, G.A. (2008) Photo-control of peptide conformation on a timescale of seconds with a conformationally constrained, blue-absorbing, photo-switchable linker. *Org. Biomol. Chem.*, **6**, 4323–4332.

16 Chi, L., Sadovski, O., and Woolley, G.A. (2006) A blue-green absorbing cross-linker for rapid photoswitching of peptide helix content. *Bioconjugate Chem.*, **17**, 670–676.

17 James, D.A., Burns, D.C., and Woolley, G.A. (2001) Kinetic characterization of ribonuclease S mutants containing photoisomerizable phenylazophenylalanine residues. *Protein Eng.*, **14**, 983–991.

18 Cordes, T., Elsner, C., Herzog, T.T., Hoppmann, C., Schadendorf, T., Summerer, W., Ruck-Braun, K., and Zinth, W. (2009) Ultrafast Hemithioindigo-based peptide-switches. *Chem. Phys.*, **358**, 103–110.

19 Cordes, T., Heinz, B., Regner, N., Hoppmann, C., Schrader, T.E., Summerer, W., Ruck-Braun, K., and Zinth, W. (2007) Photochemical Z -> E isomerization of a hemithioindigo/hemistilbene omega-amino acid. *ChemPhysChem.*, **8**, 1713–1721.

20 Priewisch, B., Steinle, W., and Ruck-Braun, K. (2005) Novel in photoswitchable amino acids, *Peptides 2004: Proceedings of the Third International and Twenty-Eighth European Peptide Symposium*, (eds M. Flegel, M. Fridkin, C. Gilon, and J. Slaninova) Kenes International, Geneva. 8pp. 756–757.

21 Guerrero, L., Smart, O.S., Weston, C.J., Burns, D.C., Woolley, G.A., and Allemann, R.K. (2005) Photochemical regulation of DNA-binding specificity of MyoD. *Angew Chem. Int. Ed. Engl.*, **44**, 7778–7782.

22 Nomura, A. and Okamoto, A. (2009) Photoresponsive tandem zinc finger peptide. *Chem. Commun.*, 1906–1908.

23 Woolley, G.A., Jaikaran, A.S., Berezovski, M., Calarco, J.P., Krylov, S.N., Smart, O.S., and Kumita, J.R. (2006) Reversible photocontrol of DNA binding by a designed GCN4-bZIP protein. *Biochemistry*, **45**, 6075–6084.

24 Kneissl, S., Loveridge, E.J., Williams, C., Crump, M.P., and Allemann, R.K. (2008) Photocontrollable peptide-based switches target the anti-apoptotic protein Bcl-xL. *ChemBioChem*, **9**, 3046–3054.

25 Parisot, J., Kurz, K., Hilbrig, F., and Freitag, R. (2009) Use of azobenzene amino acids as photo-responsive conformational switches to regulate antibody-antigen interaction. *J. Sep. Sci.*, **32**, 1613–1624.

26 Kim, Y., Phillips, J.A., Liu, H., Kang, H., and Tan, W. (2009) Using photons to manipulate enzyme inhibition by an azobenzene-modified nucleic acid probe. *Proc. Natl. Acad. Sci. USA*, **106**, 6489–6494.

27 Hayashi, G., Hagihara, M., and Nakatani, K. (2009) RNA aptamers that reversibly bind photoresponsive azobenzene-containing peptides. *Chemistry*, **15**, 424–432.

28 Woolley, G.A. (2005) Photocontrolling peptide alpha helices. *Acc. Chem. Res.*, **38**, 486–493.

29 Zhang, F., Zarrine-Afsar, A., Al-Abdul-Wahid, M.S., Prosser, R.S.,

Davidson, A.R., and Woolley, G.A. (2009) Structure-based approach to the photocontrol of protein folding. *J. Am. Chem. Soc.*, **131**, 2283–2289.

30 Renner, C. and Moroder, L. (2006) Azobenzene as conformational switch in model peptides. *ChemBioChem.*, **7**, 868–878.

31 Mayer, G. and Heckel, A. (2006) Biologically active molecules with a "light switch". *Angew Chem. Int. Ed. Engl.*, **45**, 4900–4921.

32 Young, D.D. and Deiters, A. (2007) Photochemical control of biological processes. *Org. Biomol. Chem.*, **5**, 999–1005.

33 Edenhofer, F. (2008) Protein transduction revisited: novel insights into the mechanism underlying intracellular delivery of proteins. *Curr. Pharm. Des.*, **14**, 3628–3636.

34 Graziadei, L., Burfeind, P., and Bar-Sagi, D. (1991) Introduction of unlabeled proteins into living cells by electroporation and isolation of viable protein-loaded cells using dextran-fluorescein isothiocyanate as a marker for protein uptake. *Anal. Biochem.*, **194**, 198–203.

35 McNeil, P.L. (2001) Direct introduction of molecules into cells. *Curr. Protoc. Cell Biol.*, Chapter 20, Unit 20.1.

36 Muranaka, N., Hohsaka, T., and Sisido, M. (2002) Photoswitching of peroxidase activity by position-specific incorporation of a photoisomerizable non-natural amino acid into horseradish peroxidase. *FEBS Lett.*, **510**, 10–12.

37 Bose, M., Groff, D., Xie, J., Brustad, E., and Schultz, P.G. (2006) The incorporation of a photoisomerizable amino acid into proteins in *E. coli. J. Am. Chem. Soc.*, **128**, 388–389.

38 Gurney, A.M. and Lester, H.A. (1987) Light-flash physiology with synthetic photosensitive compounds. *Physiol. Rev.*, **67**, 583–617.

39 Gorostiza, P. and Isacoff, E. (2007) Optical switches and triggers for the manipulation of ion channels and pores. *Mol. Biosyst.*, **3**, 686–704.

40 Gorostiza, P. and Isacoff, E.Y. (2008) Optical switches for remote and noninvasive control of cell signaling. *Science*, **322**, 395–399.

41 Kramer, R.H., Chambers, J.J., and Trauner, D. (2005) Photochemical tools for remote control of ion channels in excitable cells. *Nat. Chem. Biol.*, **1**, 360–365.

42 Dias, A.R., Minas da Piedade, M.E., Martinho Simoes, J.A., Simoni, J.A., Teixeira, C., Diogo, H.P., Meng-Yan, Y., and Pilcher, G. (1992) Enthalpies of formation of cis-azobenzene and trans azobenzene. *J. Chem. Thermodyn.*, **24**, 439–447.

43 Rau, H. (1990) Photoisomerization of azobenzenes, in *Photochemistry and Photophysics* (ed. J.F. Rabek), CRC Press Inc., Boca Raton, FL, pp. 119–141.

44 Sadovski, O., Beharry, A.A., Zhang, F., and Woolley, G.A. (2009) Spectral tuning of azobenzene photoswitches for biological applications. *Angew Chem. Int. Ed. Engl.*, **48**, 1484–1486.

45 Borisenko, V., Burns, D.C., Zhang, Z.H., and Woolley, G.A. (2000) Optical switching of ion-dipole interactions in a gramicidin channel analogue. *J. Am. Chem. Soc.*, **122**, 6364–6370.

46 Erlanger, B.F. (1976) Photoregulation of biologically active macromolecules. *Annu. Rev. Biochem.*, **45**, 257–283.

47 Bieth, J., Wassermann, N., Vratsanos, S.M., and Erlanger, B.F. (1970) Photoregulation of biological activity by photochromic reagents, IV. A model for diurnal variation of enzymic activity. *Proc. Natl. Acad. Sci. USA*, **66**, 850–854.

48 Deal, W.J., Erlanger, B.F., and Nachmansohn, D. (1969) Photoregulation of biological activity by photochromic reagents. 3. Photoregulation of bioelectricity by acetylcholine receptor inhibitors. *Proc. Natl. Acad. Sci. USA*, **64**, 1230–1234.

49 Nerbonne, J.M., Sheridan, R.E., Chabala, L.D., and Lester, H.A. (1983) cis-3,3'-Bis-[alpha-(trimethylammonium) methyl]azobenzene (cis-Bis-Q). Purification and properties at acetylcholine receptors of electrophorus electroplaques. *Mol. Pharmacol.*, **23**, 344–349.

50 Volgraf, M., Gorostiza, P., Szobota, S., Helix, M.R., Isacoff, E.Y., and Trauner, D. (2007) Reversibly caged glutamate: a photochromic agonist of ionotropic glutamate receptors. *J. Am. Chem. Soc.*, **129**, 260–261.

51 Bartels, E., Wassermann, N.H., and Erlanger, B.F. (1971) Photochromic activators of the acetylcholine receptor. *Proc. Natl. Acad. Sci. USA*, **68**, 1820–1823.

52 Banghart, M., Borges, K., Isacoff, E., Trauner, D., and Kramer, R.H. (2004) Light-activated ion channels for remote control of neuronal firing. *Nat. Neurosci.*, **7**, 1381–1386.

53 Chambers, J.J., Banghart, M.R., Trauner, D., and Kramer, R.H. (2006) Light-induced depolarization of neurons using a modified Shaker K(+) channel and a molecular photoswitch. *J. Neurophysiol.*, **96**, 2792–2796.

54 Volgraf, M., Gorostiza, P., Numano, R., Kramer, R.H., Isacoff, E.Y., and Trauner, D. (2006) Allosteric control of an ionotropic glutamate receptor with an optical switch. *Nat. Chem. Biol.*, **2**, 47–52.

55 Fortin, D.L., Banghart, M.R., Dunn, T.W., Borges, K., Wagenaar, D.A., Gaudry, Q., Karakossian, M.H., Otis, T.S., Kristan, W.B., Trauner, D., and Kramer, R.H. (2008) Photochemical control of endogenous ion channels and cellular excitability. *Nat. Methods*, **5**, 331–338.

56 Cravatt, B.F., Wright, A.T., and Kozarich, J.W. (2008) Activity-based protein profiling: from enzyme chemistry to proteomic chemistry. *Annu. Rev. Biochem.*, **77**, 383–414.

57 Boulegue, C., Loweneck, M., Renner, C., and Moroder, L. (2007) Redox potential of azobenzene as an amino acid residue in peptides. *ChemBioChem.*, **8**, 591–594.

58 Chen, P.R., Groff, D., Guo, J., Ou, W., Cellitti, S., Geierstanger, B.H., and Schultz, P.G. (2009) A facile system for encoding unnatural amino acids in mammalian cells. *Angew Chem. Int. Ed. Engl.*, **48**, 4052–4055.

59 Knopfel, T. (2008) Expanding the toolbox for remote control of neuronal circuits. *Nat. Methods*, **5**, 293–295.

60 Yasuda, R. and Augustine, G.J. (2008) Optogenetic probes. *Brain Cell Biol.*, **36**, 1–2.

61 Palczewski, K. (2006) G protein-coupled receptor rhodopsin. *Annu. Rev. Biochem.*, **75**, 743–767.

62 Sineshchekov, O.A., Govorunova, E.G., and Spudich, J.L. (2009) Photosensory functions of channelrhodopsins in native algal cells. *Photochem. Photobiol.*, **85**, 556–563.

63 Nagel, G., Ollig, D., Fuhrmann, M., Kateriya, S., Musti, A.M., Bamberg, E., and Hegemann, P. (2002) Channelrhodopsin-1: a light-gated proton channel in green algae. *Science*, **296**, 2395–2398.

64 Nagel, G., Szellas, T., Huhn, W., Kateriya, S., Adeishvili, N., Berthold, P., Ollig, D., Hegemann, P., and Bamberg, E. (2003) Channelrhodopsin-2, a directly light-gated cation-selective membrane channel. *Proc. Natl. Acad. Sci. USA*, **100**, 13940–13945.

65 Evanko, D. (2007) Optical excitation yin and yang. *Nat. Methods*, **4**, 384.

66 Liewald, J.F., Brauner, M., Stephens, G.J., Bouhours, M., Schultheis, C., Zhen, M., and Gottschalk, A. (2008) Optogenetic analysis of synaptic function. *Nat. Methods*, **5**, 895–902.

67 Zhang, F., Aravanis, A.M., Adamantidis, A., de Lecea, L., and Deisseroth, K. (2007) Circuit-breakers: optical technologies for probing neural signals and systems. *Nat. Rev. Neurosci.*, **8**, 577–581.

68 Han, X., Qian, X., Bernstein, J.G., Zhou, H.H., Franzesi, G.T., Stern, P., Bronson, R.T., Graybiel, A.M., Desimone, R., and Boyden, E.S. (2009) Millisecond-timescale optical control of neural dynamics in the nonhuman primate brain. *Neuron*, **62**, 191–198.

69 Bi, A., Cui, J., Ma, Y.P., Olshevskaya, E., Pu, M., Dizhoor, A.M., and Pan, Z.H. (2006) Ectopic expression of a microbial-type rhodopsin restores visual responses in mice with photoreceptor degeneration. *Neuron*, **50**, 23–33.

70 Lagali, P.S., Balya, D., Awatramani, G.B., Munch, T.A., Kim, D.S., Busskamp, V.,

Cepko, C.L., and Roska, B. (2008) Light-activated channels targeted to ON bipolar cells restore visual function in retinal degeneration. *Nat. Neurosci.*, **11**, 667–675.

71 Zhao, S., Cunha, C., Zhang, F., Liu, Q., Gloss, B., Deisseroth, K., Augustine, G.J., and Feng, G. (2008) Improved expression of halorhodopsin for light-induced silencing of neuronal activity. *Brain Cell Biol.*, **36**, 141–154.

72 Gradinaru, V., Thompson, K.R., and Deisseroth, K. (2008) eNpHR: a Natronomonas halorhodopsin enhanced for optogenetic applications. *Brain Cell Biol.*, **36**, 129–139.

73 Lin, J.Y., Lin, M.Z., Steinbach, P., and Tsien, R.Y. (2009) Characterization of engineered channelrhodopsin variants with improved properties and kinetics. *Biophys. J.*, **96**, 1803–1814.

74 Zhang, F., Prigge, M., Beyriere, F., Tsunoda, S.P., Mattis, J., Yizhar, O., Hegemann, P., and Deisseroth, K. (2008) Red-shifted optogenetic excitation: a tool for fast neural control derived from Volvox carteri. *Nat. Neurosci.*, **11**, 631–633.

75 Airan, R.D., Thompson, K.R., Fenno, L.E., Bernstein, H., and Deisseroth, K. (2009) Temporally precise in vivo control of intracellular signalling. *Nature*, **458**, 1025–1029.

76 Iseki, M., Matsunaga, S., Murakami, A., Ohno, K., Shiga, K., Yoshida, K., Sugai, M., Takahashi, T., Hori, T., and Watanabe, M. (2002) A blue-light-activated adenylyl cyclase mediates photoavoidance in Euglena gracilis. *Nature*, **415**, 1047–1051.

77 Schroder-Lang, S., Schwarzel, M., Seifert, R., Strunker, T., Kateriya, S., Looser, J., Watanabe, M., Kaupp, U.B., Hegemann, P., and Nagel, G. (2007) Fast manipulation of cellular cAMP level by light *in vivo*. *Nat. Methods*, **4**, 39–42.

78 Rockwell, N.C., Su, Y.S., and Lagarias, J.C. (2006) Phytochrome structure and signaling mechanisms. *Annu. Rev. Plant. Biol.*, **57**, 837–858.

79 Shimizu-Sato, S., Huq, E., Tepperman, J.M., and Quail, P.H. (2002) A light-switchable gene promoter system. *Nat. Biotechnol.*, **20**, 1041–1044.

80 Mendelsohn, A.R. (2002) An enlightened genetic switch. *Nat. Biotechnol.*, **20**, 985–987.

81 Tyszkiewicz, A.B. and Muir, T.W. (2008) Activation of protein splicing with light in yeast. *Nat. Methods*, **5**, 303–305.

82 Mootz, H.D., Blum, E.S., Tyszkiewicz, A.B., and Muir, T.W. (2003) Conditional protein splicing: a new tool to control protein structure and function in vitro and in vivo. *J. Am. Chem. Soc.*, **125**, 10561–10569.

83 Leung, D.W., Otomo, C., Chory, J., and Rosen, M.K. (2008) Genetically encoded photoswitching of actin assembly through the Cdc42-WASP-Arp2/3 complex pathway. *Proc. Natl. Acad. Sci. USA*, **105**, 12797–12802.

84 Strickland, D., Moffat, K., and Sosnick, T.R. (2008) Light-activated DNA binding in a designed allosteric protein. *Proc. Natl. Acad. Sci. USA*, **105**, 10709–10714.

85 Moglich, A., Ayers, R.A., and Moffat, K. (2009) Design and signaling mechanism of light-regulated histidine kinases. *J. Mol. Biol.*, **385**, 1433–1444.

86 Nienhaus, G.U. and Wiedenmann, J. (2009) Structure, dynamics and optical properties of fluorescent proteins: perspectives for marker development. *ChemPhysChem.*, **10** (9-10), 1369–1379.

6
Optical Stimulation of Neurons

S.M. Rajguru, A.I. Matic, and C.-P. Richter

6.1
Introduction

Motivations to stimulate neurons include research that studies normal function of nerves and sensory systems, with the ultimate goal to design devices that are able to restore or replace neural function. Some contemporary neural prostheses restore motor function in patients who have suffered from stroke, cochlear implants restore hearing in severe-to-profound deaf individuals, vestibular prostheses restore balance and treat vertigo, and deep brain stimulators aim to decrease the symptoms of Parkinson's disease. However, the devices stimulate the neurons with electrical current, which spreads in tissue and renders spatially selective stimulation difficult. Electrical stimuli also produce stimulation artifacts, making stimulation and simultaneous recording of the neural response challenging. Furthermore, for electrical stimulation the electrode must have direct contact with the tissue, which can result in tissue damage caused by the physical contact of the electrode or by large current densities generated close to the electrode. It has recently been suggested that many of these limitations may be overcome by using pulsed near-infrared (NIR) radiation from a laser for stimulation.

The idea of exciting cells using light dates back to 1891. The first reports of photostimulation of neural tissue were published as early as 1947. Arvanitaki and Chalazonitis [1] demonstrated that radiation at different wavelengths could produce effects in the visceral ganglia of *Aplysia*. Following Arvanitaki and Chalazonitis, different approaches have been used to irradiate and stimulate neurons. One particular radiation sources was the laser [2–10]. Lasers have been used to study the local dynamic responses of cultured neurons, or to identify functional connections in neural networks [11–13]. More recent advances in laser technology have also opened up the possibility for laser use in the field of neuroprostheses [5, 14, 15]. Despite the previous reports on using lasers to stimulate neurons, it is difficult to compare the results from early experiments and consider them as a logical progression towards the present method, because

Laser Imaging and Manipulation in Cell Biology. Edited by Francesco S. Pavone
Copyright © 2010 WILEY-VCH Verlag GmbH & Co. KGaA, Weinheim
ISBN: 978-3-527-40929-7

the optical parameters used in early experiments differ significantly from those used here.

The use of pulsed near-infrared lasers for neural stimulation has several advantages over electric current. The radiation can be delivered in a non-contact manner, which eliminates the need for direct contact with the tissue to be stimulated. Moreover, laser stimulation produces artifact-free responses from the neurons, allowing adjacent placement of stimulating and recording electrodes. Lasers can also be focused with very fine resolution and achieve higher spatially selectivity since near-infrared radiation does not spread in the tissue as much as electrical current. Recently, it has also been shown that continuous stimulation with pulsed infrared lasers over durations of 8 h does not result in any significant tissue damage [5, 16]. These advantages have led to a substantial effort towards developing various optical stimulation techniques. In this chapter, we will discuss the use of lasers in stimulation of neural tissue by presenting a brief review of the work that has been accomplished, and discuss recent advances in our understanding of pulsed mid-infrared lasers and the challenges that need to be addressed in the future.

6.2
Neural Stimulation with Optical Radiation

6.2.1
General Considerations

The idea and subsequent efforts to use optical radiation for stimulating neurons is not novel. With mixed results, many groups have studied the effects of radiation at different wavelengths on neurons. Despite reports in the literature of successful approaches to stimulate neurons with optical radiation, it is difficult to provide a direct comparison of results obtained in earlier studies and the current use of lasers in neurostimulation. During the pre-laser time, radiation sources were bright lamps and optical neural modulation was explained by the interaction between the radiation with either naturally existing chromophores in the cell or by the interaction between the radiation and artificially induced chromophores. With the development of lasers, a novel approach was possible. However, neural stimulation was explained by either the interaction between the laser radiation of a given wavelength and a chromophore or by the direct action on the cell membrane resulting in transient holes. The recent approach differs from earlier approaches because interactions with chromophores or the formation of transient holes in the cell membrane do not explain the laser–tissue interaction that leads to action potentials of the neurons. Parameters, including the radiation wavelength, time of irradiation, power, as well as the absorbing and scattering characteristics of the target structure are different in the recent experiments and determine the effect of the irradiation.

6.2.2
Effect of Optical Stimulation on Excitability

Optical stimulation has been used concurrently with electrical stimulation to modulate electrically evoked neural activity. Studies by Booth *et al.* [17] and Cook *et al.* [18] have found an increase in electrical stimulation threshold and a decrease in the electrically evoked action potential amplitude when irradiating isolated nerve fibers with continuous, ultraviolet light. Since irradiation modulated neural activities only for radiation wavelengths below 320 nm, the authors concluded that a photochemical effect on an unknown membrane substance was responsible for the results. Olson *et al.* [6, 7] have studied the changes in excitability of nerve cells from rat cerebellum by irradiating the tissue with a pulsed ruby laser or a pulsed dye laser concurrent with electric current pulses. They found that low energy pulses produced no noticeable effect, while at high energies the cells did not respond to electrical stimulation. They also observed mitochondrial damage in electron micrographs when high energies were used. The results were similar for laser wavelengths between 490 and 685 nm. The authors excluded the possibility of a photochemical reaction based on their observations that sensitivity at 690 nm depended on the duration of laser pulses. A purely photochemical mechanism would depend only on the total energy of the pulse but not its rate or duration. They concluded that the irradiation induced a local heating in the mitochondria, which led to tissue damage and hence the changes in excitability. Uzdensky and Savransky [19] have demonstrated an initial increase and a subsequent decrease in the frequency of electrically evoked action potentials when irradiating neurons with a dye laser. Wesselmann *et al.* have utilized a pulsed Nd:YAG laser to modify the electrically evoked activity from spinal columns [20] as well as dorsal roots and peripheral nerves [21]. In both studies, they measured a reduction in amplitude of several peaks of the compound action potential and significant increases in tissue temperature during irradiation, up to 60 °C. In a follow-up histological study, they measured an increase in nerve fiber diameter and a change in cytoskeletal content in irradiated nerves [22]. Another group has investigated the effects of Nd:YAG irradiation on excised spinal nerves and measured a laser-induced decrease in the electrically evoked action potential in a dose-dependent fashion [23]. Bagis *et al.* [24, 25] on the other hand, have shown that low-energy Ga-Ar irradiation had no influence on electrical stimulation threshold, action potential latency or duration and did not change nerve excitability.

6.2.3
Optical Stimulation via Photochemical Mechanism

Photochemical reactions induced by lasers are well studied in the photobiology literature. In applications of photochemical reactions, laser energy is converted into a form of chemical energy by exciting a molecular chromophore using light. In its excited state, the chromophore subsequently participates in the intended chemical reaction(s). In most applications, the molecular chromophores are exogenously added. This is exercised for example during photodynamic chemotherapy [26]. Although less commonly used, photostimulation has been shown in even endogenous chromophores such as amino acids or pigments.

6.2.3.1 Activation via Exogenously Added Chromophore

Neural activation via the excitation of an exogenously added fluorophore or a "caged" molecule that is released upon irradiation has been possible. An early approach used by Farber and Grinvald [3] was to design photosensitive molecular probes that, in the presence of light, depolarized the membrane by forming transient channels. When a stained neuron was irradiated by He:Ne laser, the membrane was depolarized and action potentials were evoked. The authors used this technique to map synaptic connections of leech neurons. However, they also noted that only a limited number of repetitions with photostimulation were possible as this method often caused irreversible photochemical damage. Hirase *et al.* [12] have used a pulsed Ti:sapphire laser and were able to elicit action potentials with and without an exogenous fluorophore. In their experiments, they observed two regimes of optical stimulation, which they called Type 1 and Type 2. Type 1 stimulation occurred at laser powers of 200–400 mW. Based on the optical parameters used (750–800 nm wavelength, 300–350 mW power, and >30 ms duration), the likely underlying mechanism involved a two-photon reaction with an endogenous chromophore that could have been assisted by the exogenously applied fluorophore. When using the exogenous fluorophore alone, they could not stimulate the neurons when irradiating with a wavelength longer than 800 nm. However, stimulation occurred all times when irradiating with wavelengths 790 nm or less. Type 2 stimulation was achieved at higher radiation intensities. The authors hypothesized that the second type of stimulation could have formed microholes in the neuron plasma membrane, which allowed an ion flux into the cell and subsequent neural depolarization.

In 1990, Wilcox *et al.* reported the synthesis of light-sensitive precursors to neurotransmitters [27]. The molecules are inert initially, but when irradiated with ultraviolet light they become active and can bind with their respective neuroreceptors. Dalva and Katz [28, 29] used this technique to map the developing synaptic connections in the primary visual cortex of ferrets by uncaging photolabile glutamate. Since then, many investigators have taken advantage of these photolabile neurotransmitter precursors, most notable glutamate, in their research (for reviews, see Callaway and Yuste [30] and Eder *et al.* [31]).

6.2.3.2 Activation of an Endogenous Chromophore

Utilizing endogenous chromophores for a photochemical reaction is less common; however, some research groups report photostimulation. Early studies were carried out by Arvanitaki and Chalazonitis [32] and Fork [33]. Each of these researchers irradiated photosensitive *in vitro* neural preparations with continuous or quasi-continuous light (duration of pulses is significantly long, such that no confinement of the optical energy occurs in the tissue) and measured the rate of action potential generation in response to irradiation. Fork [33] mapped the interconnections of *Aplysia* neurons without causing damage, but was unable to determine the mechanism of the optical stimulation. The author used ion-free solutions and blocking agents to study the mechanism of light-induced stimulation. The results suggested that a Na^+ ion channel is involved in the process. However, the results were found to be inconclusive since some response persisted even with the addition of ouabain. Arvanitaki and Chalazonitis [32] used several preparations, such as *Aplysia* and a stained *Sepia* giant axon, and irradiated with visible (400–700 nm) and infrared (750–4000 nm) light. They found that most neurons changed their excitation patterns in response to irradiation as determined by an increase in spontaneous rate or initiation of depolarization in non-spontaneously active neurons. While for most of the neurons the irradiation resulted in an increase in neural activities (excitation), one type of nerve cell in an *Aplysia* preparation showed inhibition when irradiated with infrared light. The authors suggested that the findings could be explained by the absorption of light by an undetermined cellular molecule, leading to either neural depolarization or neural inhibition. More recently, Lima and Miesenbock [34] have controlled behavioral responses in *Drosophila* by utilizing an endogenous rhodopsin-gated ion channel, the synthesis of which is described by Zemelman *et al.* [35, 36]. Further advances in genetically induced photostimulation implemented an algal channelrhodopsin [37] to improve the temporal response of the neurons to the activating radiation [38] (for a review, see Herlitze *et al.* [39]). The chromophore channelrhodopsin-2 has been delivered to neurons of rat hippocampal cell cultures by using a lentiviral delivery system. After the protein has been expressed in the neurons, the cells were activated with 500 ms blue-light pulses ($8–12\ mW\ mm^{-2}$) [40, 41]. The results were confirmed in two animal models [42–44].

6.3
Direct Optical Stimulation of Neural Tissue

More recently, laser–neuron interactions have been studied that were not attributed to photochemical reactions. Balaban *et al.* [45] have used a HeNe laser to irradiate snail ganglia. They were only able to modify the rate of action potentials in spontaneously active neurons and were not able to elicit action potentials in silent neurons. The authors noted that the temperature effects of irradiation were responsible for the outcomes of the experiments. A recent study showed that irradiation with a pulsed, near-infrared diode laser ($\lambda = 980$ nm and 4 ms to 400 ms pulse length) induced inward currents in dissociated cell cultures of nociceptive rat dorsal root ganglion cells [46]. At the stimulation threshold fluence of $2.8\ J\ cm^{-2}$, they calculated a temperature of 42 °C at the neurons. Interestingly, the radiation induced inward current decreased in amplitude with successive irradiation attempts of the same neurons using the same optical parameters. The authors suggested that the primary mechanism of stimulation was a thermal effect from the laser. Furthermore, they demonstrated that all of the laser-stimulated neurons were sensitive to capsaicin, which is bound by a temperature-sensitive ion channel. Another group found that neurons in a thalamic nucleus were depolarized by the irradiation of a XeCl laser ($\lambda = 308$ nm; 40 ns pulse duration; 1 Hz repetition rate). However, radiant exposures required to evoke a response $(0.9\,\mathrm{J\,cm^{-2}})$ were very near the ablation threshold $(1.0\,\mathrm{J\,cm^{-2}})$ [47]. Though not a direct measure of the electrical activity of neurons, Wade *et al.* [48] have measured an increase in GABA neurotransmitter release upon irradiating with low levels of continuous white light, and a suppression of GABA release with higher levels of irradiation.

6.3.1
Pulsed Infrared Lasers for Direct Stimulation

A new paradigm of neural stimulation using optical means has been developed over the last 5 years. Pulses of near-infrared radiation are delivered, which depolarizes the neurons in the optical path in a temporally correlated manner. This method of neural stimulation does not require contact between the optical source and the neural tissue, thereby allowing radiation delivery in air. Most of the radiation wavelengths used for these studies (1.84–2.12 μm) can be delivered via optical fiber, which provides flexibility of minimally invasive delivery. Furthermore, simultaneous optical stimulation and electrical recording is simple, as there is no stimulation artifact present in the data. The most attractive feature of optical neural stimulation is the improvement in spatial selectivity of stimulation. The optical radiation does not spread significantly in the tissue, compared to electric current, and can be further focused using lenses. The ability to confine the optical stimulus to a smaller portion of neurons may allow for more discrete neural stimulation with improved resolution. While there are many esearch questions still to answer regarding optical stimulation, the initial results are promising. Below, we present the research to date on pulsed, mid-infrared stimulation of neurons.

6.3.1.1 Stimulation of Peripheral Nerves

Wells *et al.* [15, 49, 50] have used a pulsed infrared laser to stimulate the sciatic nerve and have pioneered a new field of optical radiation to evoke neural activity. They successfully evoked action potentials from the rat sciatic nerve in response to laser pulses without the use of exogenous fluorophores or genetically modified neurons. The stimulation source for their studies was a free electron laser (FEL, $\lambda = 2$–6 μm; $\tau_p = 5\,\mu s$), a Ho:YAG laser ($\lambda = 2.12$ μm; $\tau_p = 350\,\mu s$) at low energies, or a pulsed diode laser ($\lambda = 1860$ nm; $\tau_p = 10\,\mu s$ to 10 ms; Lockheed Martin Aculight (LMA)). It has been demonstrated that neural stimulation is possible at optical wavelengths that are moderately absorbed in neural tissue. The estimated optical penetration depth in neural tissue is 200–1500 μm [51, 52], assuming primarily water absorption [53, 54]. The irradiation of different portions of the sciatic nerve was found to elicit contraction in different muscle groups and it has been hypothesized that the mechanism of stimulation is local, transient temperature increase from light absorption by water in the tissue [55]. In another series of experiments, rat cavernous nerves were successfully stimulated with a pulsed laser at a near-infrared radiation wavelength [56, 57].

Wells *et al.* [58, 59] have also explored whether the irradiation of the nerve with near-infrared radiation results in nerve damage by stimulating rat sciatic nerves with the Ho:YAG laser. They observed that the upper limit for safe laser stimulation repetition rate in peripheral nerves occurs near 5 Hz and that maximum duration for constant low repetition rate stimulation (2 Hz) is approximately 4 min with adequate tissue hydration. Notably, this is different from the findings for the auditory nerve for which stimulation rates of 200 Hz over 8 h did not result in noticeable functional damages (see below for details).

6.3.1.2 Stimulation of Cranial Nerves

Based on the experiments by Wells, optical radiation has been used to stimulate the facial [60] and auditory nerves [5, 61–63]. Teudt et al. [60] have irradiated the facial nerve using a 600 μm optical fiber with exposures between 0.71 and 1.77 J cm^{-2} and measured evoked compound muscle action potentials (CmAPs) at the facial muscles. Optical stimulation of the auditory nerve has been studied in normal hearing gerbils [64–66] and deafened gerbils [67], the mouse [68], guinea pig [69] and cat [16]. In a first series of experiments a Ho:YAG laser was used to stimulate the auditory nerve of gerbils [5]. For neural stimulation, the laser was coupled to a 200 μm-diameter optical fiber, which was placed in front of the cochlear round window in gerbils. The radiation beam was directed towards the spiral ganglion cells and compound action potentials were evoked directly in response to the laser pulse. An increase in laser energy induced a monotonic increase in the evoked response. Subsequent experiment with different LMA diode lasers ($\lambda = 1.844$–1.94 μm; $\tau_p = 5$ μs–1 ms; $f = 2$–400 Hz) showed that the patterns of the compound action potentials changed with increasing pulse length [62, 63]. In acutely and chronically deaf gerbils, thresholds for optically evoked CAPs were not significantly elevated when stimulating with short pulse durations [67]. Long-term acute stimulation was achieved for 6 h in the gerbils at 13 Hz [65] and 8 h in cats at 200 Hz [16] and showed stable CAP amplitudes for the duration of stimulation. Selectivity of stimulation in the auditory system of rodents has been investigated with several methods. The results indicate that a single channel optical stimulus is more selective than a monopolar electrical stimulus (the type of stimulus most frequently used in cochlear implants) [70] and is on the order of selectivity of a low level acoustic tone [69, 71]. Experiments are underway currently to investigate the safety and efficacy of a chronic optical cochlear implant. These data will provide information on the maximum non-damaging repetition rate of stimulation. It remains to be seen if the neural responses evoked by the optical stimulus correlate into a meaningful perception of sound.

Another potential application of pulsed infrared lasers has been suggested in vestibular prostheses, as a means to reduce imbalance and disorientation caused by vestibular dysfunction. Harris and coworkers [72] used optical stimulation with 1.84 μm radiation ($\tau_p = 10$ μs – 1 ms) to irradiate vestibular nerve. However, optical stimulation of the ampullae did not evoke detectable eye movements. In a different set of experiments, Bradley et al. have stimulated Scarpa's ganglion in gerbils, while recording neural activities from single vestibular nerve fibers. The experiments were inconclusive. With the methods available, it was not possible to study any evoked eye movements. Recently, Rajguru et al. [73] have showed that inner-ear hair cells are exquisitely sensitive to infrared radiation at 1862 nm, and modulate neurotransmitter release in response to short IR pulses through an endogenous mechanism. The transmitter release from hair cells was monitored in the crista ampullaris by recording post-synaptic afferent responses. Some afferents fired an action potential in response to each IR pulse delivered to hair cells, at rates up to 96 pulses per second. In addition to phase-locking, profound tonic changes in afferent discharge rate were evoked by continuous IR pulse trains. The data suggest that pulsed IR radiation activates hair cell intracellular signaling, presumably $Ca2+$, leading to modulation of synaptic transmission.

6.3.1.3 Advantages of Optical Stimulation

The use of lasers in stimulation of neurons and other excitable cells is of increasing interest, especially given the success of pulsed and continuous lasers in exciting neurons. A potential application of the pulsed laser light is in neuroprostheses and it may be advantageous in several ways over traditional electrical stimulation. Optical stimulation yields artifact-free signals, which enables the simultaneous recording of neural responses. This is difficult with electrical stimulation and will allow for long-term evaluation of the state of the tissue. It is also advantageous that light can be delivered in a non-contact manner. The observations from gerbils and cats, where long duration stimulation did not affect the amplitude of evoked compound action potentials, lend further support to the safety of lasers in chronic implant use.

In addition, electrical stimulation may not be able to achieve the spatial selectivity of optical stimulation. The main advantage of optical nerve stimulation, its spatial selectivity, has been demonstrated for the sciatic nerve, the facial nerve, and the auditory nerve. In the inner ear, the transiently expressed transcription factor c-FOS was used to stain activated nerve cells to identify the spatial area of the cochlea that is stimulated. Immunohistochemical staining for c-FOS in the cochlea shows that optical stimulation of the cochlea is more selective compared to electrical stimulation. At high stimulus levels, stained cells were seen directly opposite to the optical fiber for optical stimulation [65]. In contrast, electrical stimulation at high levels resulted in the activation of all spiral ganglion cell in the cochlea [65]. Furthermore, masking data and recording of neural activities from the inferior colliculus measurements demonstrate that the laser can stimulate a small population of cells similar to an acoustic toneburst. In the sciatic and facial nerve studies, different nerve bundles within the main nerve trunk were individually stimulated when the radial location of the stimulation site was varied [49, 74]. This is in contrast to electric stimulation, which evoked strong and unselective responses in all muscle groups innervated by the nerve.

6.3.2
Challenges for Optical Stimulation

6.3.2.1 Mechanism of Stimulation with Optical Radiation

One of the main questions remaining unanswered regarding optical stimulation of neural tissue is the mechanism of stimulation. Laser–tissue interactions can occur through various biophysical mechanisms, including photochemical, photothermal, photomechanical, and electric field effects. For example, Jacques [75] and Thomsen [76] have provided excellent reviews on this topic, which is further discussed in detail in Chapter 7 by Valery Tuchin. Wells and coworkers [77, 78] have investigated the various biophysical mechanisms that could be responsible for optical stimulation. The results from their studies eliminate any photochemical and electric field effects as the possible means. The individual photon energies emitted by these optical stimulation lasers are significantly lower than the energies required to move an electron to an excited state, as is needed for a photochemical reaction. Furthermore, there was no wavelength of significant enhancement for the optical stimulation, which would have suggested a photochemical mechanism. For photomechanical effects, stress waves, volumetric tissue expansion, and thermoelastic expansion were considered as mechanisms for the stimulation. Wells *et al.* [77, 78] have proposed that photomechanical effects from stress wave generation and thermoelastic expansion can be eliminated based on the finding that stimulation threshold for an action potential was independent of varying laser pulse length (all tested at a nearly constant wavelength or depth of penetration). Furthermore, all pulse lengths tested evoked an action potential.

Experimental results for the peripheral nerve and the auditory system differ. For the deafened cochlea, shorter pulse lengths are apparently more efficient in evoking an action potential. This conclusion was made from graphs showing the peak-to-peak amplitude of the action potential versus the radiation energy or radiation exposure. However, if the same compound action potential amplitudes are plotted versus the peak radiation power, shorter pulse lengths no longer appear more efficient. In other words, it appears that the transient of the optical pulse is the important factor and not the total energy.

From the Wells *et al.* study [79], photomechanical effects from volumetric expansion could not be altogether excluded since a small, but not negligible, displacement of tissue did occur at radiant exposures slightly above threshold. However, the introduction of a pressure transient having the same temporal characteristics as the optical pulse failed to induce neural depolarization at $30\times$ optical stimulation threshold. A photothermal mechanism was concluded to be the most likely means of depolarizing neurons via infrared nerve stimulation. Wells *et al.* also concluded that a temporal and spatial thermal confinement is necessary to achieve optical stimulation. Thermal confinement exists when the thermalized optical energy delivered by a single pulse accumulates in the irradiated tissue before any of the heat can dissipate through conduction or convection. For thermal confinement to be achieved at these wavelengths, the pulse length of the stimuli are typically longer than 500 ns and shorter than 200 ms. When considering a photothermal or heat-induced mechanism for neural stimulation with the pulsed infrared laser, the primary mechanism of stimulation is that the radiant energy is absorbed by neural tissue and transformed into heat, which then mediates a secondary, yet unknown, mechanism of ion flux through the neural membrane. Potential mechanistic candidates include thermal activation of a particular ion channel, thermally induced biophysical changes in the membrane (increase in ion channel conductance, decrease in Nernst equilibrium potential), or through general expansion of the neural membrane that facilitates the flux of ions. One possibility for the secondary mechanism is the activation of thermally-gated ion channels, for instance TRPV channels because it has been demonstrated that the channels can be activated by heat [80]. It has been shown that TRPV1 ion channels are expressed in the dorsal root and trigeminal ganglia of rats [80], as well as in cochlear structures of both the rat and guinea pig [81–83]. Our own immunostaining data show that the TRPV1 channel is present in gerbil and mouse spiral ganglion cells. TRPV channels are members of the transient receptor potential (TRP) superfamily. TRPV1 is perhaps most well known for being activated by the chemical capsaicin, the main ingredient in hot chili peppers, which produces a burning sensation. TRPV1 channels are also stimulated by other vanilloid compounds, acid (pH ≤ 5.9), and heat ($\geq 43\,°C$), making it a key channel in peripheral nociception. Experiments were conducted in gene-modified animals to determine the mechanism by which optical energy activates neural tissue. Most of the TRPV1 knockout mice show no auditory brain response to stimulation with optical radiation, while all control animals did. In addition to the elevation of the ABR thresholds for optical radiation, the thresholds for acoustical stimulation were elevated as well. TRPV channels apparently are important for normal auditory function. Moreover, preliminary *in vivo* experiments in gerbil showed that the infusion of the TRPV1 antagonist capsazepine in the cochlea reversibly blocks the excitation of optically evoked compound action potentials. The results provide initial evidence to support the view that the vanilloid transient receptor channel is involved in the action potential generation through near-infrared laser radiation.

6.3.2.2 Safety of Optical Stimulation

Neuroprostheses using any form of optical stimulation will be implanted in patients and expected to remain in operation over many years. Therefore, optical neural stimulation must also be optimized by establishing the parameters that allow safe, chronic stimulation of the tissue of interest without causing significant damage to neurons. To date only limited data are available for providing safe optical stimulation over long durations. Further safety studies are required to select safe optical wavelength, energy, size and placement of the optical source (fiber) for stimulating neurons.

References

1 Arvanitaki, A. and Chalazonitis, N. (1947) Reactions bioelectrques a la photoactivation des cytosomes. *Arch. Sci. Physiol. (Paris)*, **1**, 385–405.

2 Arvanitaki, A. and Chalazonitis, N. (1957) Response of neuronic soma to photoactivation of the pigmented granules of its cytoplasm. *J. Physiol. (Paris)*, **49** (1), 9–12.

3 Farber, I.C. and Grinvald, A. (1983) Identification of presynaptic neurons by laser photostimulation. *Science*, **222** (4627), 1025–1027.

4 Fork, R.L. (1971) Laser stimulation of nerve cells in aplysia. *Science*, **171** (974), 907–908.

5 Izzo, A.D. *et al.* (2006) Laser stimulation of the auditory nerve. *Lasers Surg. Med.*, **38** (8), 745–753.

6 Olson, J.E. *et al.* (1981) Laser microirradiation of cerebellar neurons in culture. Electrophysiological and morphological effects. *Cell Biophys.*, **3** (4), 349–371.

7 Olson, J.E., Schimmerling, W., and Tobias, C.A. (1981) Laser action spectrum of reduced excitability in nerve cells. *Brain Res.*, **204** (2), 436–440.

8 Rochkind, S. *et al.* (1987) Response of peripheral nerve to He-Ne laser: experimental studies. *Lasers Surg. Med.*, **7** (5), 441–443.

9 Walker, J.B. and Akhanjee, L.K. (1985) Laser-induced somatosensory evoked potentials: evidence of photosensitivity in peripheral nerves. *Brain Res.*, **344** (2), 281–285.

10 Wells, J. *et al.* (2005) Application of infrared light for *in vivo* neural stimulation. *J. Biomed. Opt.*, **10** (6), 064003.

11 Ghezzi, D. *et al.* (2008) A micro-electrode array device coupled to a laser-based system for the local stimulation of neurons by optical release of glutamate. *J. Neurosci. Methods*, **175** (1), 70–78.

12 Hirase, H. *et al.* (2002) Multiphoton stimulation of neurons. *J. Neurobiol.*, **51** (3), 237–247.

13 Hirase, H. *et al.* (2004) Calcium dynamics of cortical astrocytic networks *in vivo*. *PLoS Biol.*, **2** (4), E96.

14 Wells, J. *et al.* (2005) Optical stimulation of neural tissue *in vivo*. *Opt. Lett.*, **30** (5), 504–506.

15 Wells, J. *et al.* (2005) Application of infrared light for *in vivo* neural stimulation. *J. Biomed. Opt.*, **10**, 064003.

16 Rajguru, S.M. *et al.* (2009) Optical cochlear implants: a study of efficacy and safety in cats. Presented at Conference on Implantable Auditory Prostheses. 2009, Lake Tahoe, CA.

17 Booth, J., von, M.A., and Stampfli, R. (1950) The photochemical action of ultra-violet light on isolated single nerve fibres. *Helv. Physiol. Pharmacol. Acta*, **8** (2), 110–127.

18 Cook, J.S. (1956) Some characteristics of hemolysis by ultraviolet light. *J. Cell Physiol.*, **47** (1), 55–84.

19 Uzdensky, A.B. and Savransky, V.V. (1997) Single neuron response to pulse-periodic laser microirradiation. Action spectra and two-photon effect. *J. Photochem. Photobiol. B: Biol.*, **39**, 224–228.

20 Wesselmann, U., Lin, S.F., and Rymer, W.Z. (1991) Selective decrease of small sensory neurons in lumbar dorsal root ganglia labeled with horseradish peroxidase after ND:YAG laser irradiation of the tibial nerve in the rat. *Exp. Neurol.*, **111** (2), 251–262.

21 Wesselmann, U., Lin, S.F., and Rymer, W.Z. (1991) Effects of Q-switched Nd: YAG laser irradiation on neural impulse propagation: II. Dorsal roots and peripheral nerves. *Physiol. Chem. Phys. Med. NMR*, **23** (2), 81–100.

22 Wesselmann, U., Kerns, J.M., and Rymer, W.Z. (1994) Laser effects on myelinated and nonmyelinated fibers in the rat peroneal nerve: a quantitative ultrastructural analysis. *Exp. Neurol.*, **129** (2), 257–265.

23 Orchardson, R., Peacock, J.M., and Witters, C.J. (1997) Effect of pulsed Nd: YAG laser radiation on action potential conduction in isolated mammalian spinal nerves. *Laser Surg. Med.*, **21**, 142–148.

24 Bagis, S. *et al.* (2002) Acute electrophysiologic effect of pulsed gallium-arsenide low energy laser irradiation on configuration of compound nerve action potential and nerve excitability. *Lasers Surg. Med.*, **30** (5), 376–380.

25 Comelekoglu, U. *et al.* (2002) Electrophysiologic effect of gallium arsenide laser on frog gastrocnemius muscle. *Lasers Surg. Med.*, **30** (3), 221–226.

26 Gomer, C.J. *et al.* (1989) Properties and applications of photodynamic therapy. *Radiat. Res.*, **120** (1), 1–18.

27 Wilcox, M. *et al.* (1990) Synthesis of photolabile "precursors" of amino acid neurotransmitters. *J. Org. Chem.*, **55** (5), 1585–1589.

28 Dalva, M.B. and Katz, L.C. (1994) Rearrangements of synaptic connections in visual cortex revealed by laser photostimulation. *Science*, **265** (5169), 255–258.

29 Katz, L.C. and Dalva, M.B. (1994) Scanning laser photostimulation: a new

approach for analyzing brain circuits. *J. Neurosci. Methods*, **54** (2), 205–218.

30 Callaway, E.M. and Yuste, R. (2002) Stimulating neurons with light. *Curr. Opin. Neurobiol.*, **12** (5), 587–592.

31 Eder, M., Zieglgansberger, W., and Dodt, H.U. (2004) Shining light on neurons--elucidation of neuronal functions by photostimulation. *Rev. Neurosci.*, **15** (3), 167–183.

32 Arvanitaki, A. and Chalazonitis, N. (1961) Excitatiory and inhibitory processes initiated by light and infra-red radiations in single identifiable nerve cells, in *Nervous Inhibition* (ed. E. Florey), Pergamon Press, New York.

33 Fork, R. (1971) Laser stimulation of nerve cells in aplysia. *Science*, **171**, 907–908.

34 Lima, S.Q. and Miesenbock, G. (2005) Remote control of behavior through genetically targeted photostimulation of neurons. *Cell*, **121** (1), 141–152.

35 Zemelman, B.V. *et al.* (2002) Selective photostimulation of genetically chARGed neurons. *Neuron*, **33** (1), 15–22.

36 Zemelman, B.V. *et al.* (2003) Photochemical gating of heterologous ion channels: remote control over genetically designated populations of neurons. *Proc. Natl. Acad. Sci. USA*, **100** (3), 1352–1357.

37 Nagel, G. *et al.* (2003) Channelrhodopsin-2, a directly light-gated cation-selective membrane channel. *Proc. Natl. Acad. Sci.*, **100** (24), 13940–13945.

38 Boyden, E.S. *et al.* (2005) Millisecond-timescale, genetically targeted optical control of neural activity. *Nat. Neurosci.*, **8** (9), 1263–1268.

39 Herlitze, S. and Landmesser, L.T. (2007) New optical tools for controlling neuronal activity. *Curr. Opin. Neurobiol.*, **17** (1), 87–94.

40 Zhang, F. *et al.* (2006) Channelrhodopsin-2 and optical control of excitable cells. *Nat. Methods*, **3** (10), 785–792.

41 Boyden, E.S. *et al.* (2005) Millisecond-timescale, genetically targeted optical control of neural activity. *Nat. Neurosci.*, **8** (9), 1263–1268.

42 Wang, H. *et al.* (2007) High-speed mapping of synaptic connectivity using photostimulation in channelrhodopsin-2

transgenic mice. *Proc. Natl. Acad. Sci. USA*, **104** (19), 8143–8148.

43 Han, X. and Boyden, E.S. (2007) Multiple-color optical activation, silencing, and desynchronization of neural activity, with single-spike temporal resolution. *PLoS One*, **2** (3), e299.

44 Bernstein, J.G. *et al.* (2008) Prosthetic systems for therapeutic optical activation and silencing of genetically-targeted neurons. *Proc. Soc. Photo Opt. Instrum. Eng.*, **6854**, 68540.

45 Balaban, P. *et al.* (1992) He-Ne laser irradiation of single identified neurons. *Lasers Surg. Med.*, **12**, 329–337.

46 Greffrath, W. *et al.* (2002) Inward currents in primary nociceptive neurons of the rat and pain sensations in humans elicited by infrared diode laser pulses. *Pain*, **99** (1–2), 145–155.

47 Allegre, G., Avrillier, S., and Albe-Fessard, D. (1994) Stimulation in the rat of a nerve fiber bundle by a short UV pulse from an excimer laser. *Neurosci. Lett.*, **180**, 261–264.

48 Wade, P.D., Taylor, J., and Siekevitz, P. (1988) Mammalian cerebral cortical tissue responds to low-intensity visible light. *Proc. Natl. Acad. Sci. USA*, **85** (23), 9322–9326.

49 Wells, J.D. *et al.* (2005) Optical stimulation of neural tissue *in vivo*. *Opt. Lett.*, **30** (5), 504–506.

50 Wells, J.D. *et al.* (2005) Optical stimulation of neural tissue *in vivo*. *Opt. Lett.*, **30** (5), 504–506.

51 Falk, M. and Ford, T.A. (1966) Infrared spectrum and structure of liquid water. *Can. J. Chem.*, **44**, 1699–1707.

52 Hale, G.M. and Querry, M.R. (1973) Optical constants of water in the 200-nm to 200-μm wavelength range. *Appl. Opt.*, **12** (3), 555–563.

53 LoPachin, R.M. and Stys, P.K. (1995) Elemental composition and water content of rat optic nerve myelinated axons and glial cells: effects of *in vitro* anoxia and reoxygenation. *J. Neurosci.*, **15** (10), 6735–6746.

54 Rosenberg, H.F. and Gunner, J.T. (1959) Sodium, potassium and water contents of various components (intact, desheathed nerve, and epineurium) of

fresh (untreated) medullated (sciatic) nerve of R. ridibunda (and some comparative data on cat sciatic). *Pflügers Arch.*, **269** (3), 270–273.

55 Wells, J. *et al.* (2007) Biophysical mechanisms of transient optical stimulation of peripheral nerve. *Biophys. J.*, **93** (7), 2567–2580.

56 Fried, N.M. *et al.* (2008) Noncontact stimulation of the cavernous nerves in the rat prostate using a tunable-wavelength thulium fiber laser. *J. Endourol.*, **22** (3), 409–413.

57 Fried, N.M. *et al.* (2008) Laser stimulation of the cavernous nerves in the rat prostate, *in vivo*: optimization of wavelength, pulse energy, and pulse repetition rate. *Conf. Proc. IEEE Eng. Med. Biol. Soc.*, **2008**, 2777–2780.

58 Wells, J.D. *et al.* (2007) Optically mediated nerve stimulation: identification of injury thresholds. *Lasers Surg. Med.*, **39** (6), 513–526.

59 Wells, W.A. *et al.* (2007) Validation of novel optical imaging technologies: the pathologists' view. *J. Biomed. Opt.*, **12** (5), 051801.

60 Teudt, I.U. *et al.* (2007) Optical stimulation of the facial nerve – a new monitoring technique? *Laryngoscope*, **117**, 1641–1647.

61 Izzo, A.D. *et al.* (2007) Selectivity of neural stimulation in the auditory system: a comparison of optic and electric stimuli. *J. Biomed. Opt.*, **12** (2), 021008.

62 Izzo, A.D. *et al.* (2007) Optical parameter variability in laser nerve stimulation: a study of pulse duration, repetition rate, and wavelength. *IEEE Trans. Biomed. Eng.*, **54** (6 Pt 1), 1108–1114.

63 Izzo, A.D. *et al.* (2008) Laser stimulation of auditory neurons: effect of shorter pulse duration and penetration depth. *Biophys. J.*, **94** (8), 3159–3166.

64 Izzo, A.D. *et al.* (2006) Laser stimulation of the auditory nerve. *Laser Surg. Med.*, **38** (8), 745–753.

65 Izzo, A.D. *et al.* (2007) Optical parameter variability in laser nerve stimulation: a study of pulse duration, repetition rate, and wavelength. *IEEE Trans. Biomed. Eng.*, **54** (6 Pt 1), 1108–1114.

66 Izzo, A.D. *et al.* (2008) Laser stimulation of auditory neurons at shorter pulse

durations and penetration depths. *Biophys. J.*, **94** (8), 3159–3166.

67 Richter, C.-P. *et al.* (2008) Optical stimulation of auditory neurons: effects of acute and chronic deafening. *Hear Res.*, **242** (1–2), 42–51.

68 Suh, E. *et al.* (2009) Optical stimulation in mice lacking the TRPV1 channel. *Proc. SPIE*, **7180**, 71800S1–71800S5.

69 Rajguru, *et al.* (2010) The spread of excitation in inferior colliculus for optical cochlear stimulation. Midwinter meeting of ARO.

70 Izzo, A.D. *et al.* (2007) Selectivity of neural stimulation in the auditory system: a comparison of optic and electric stimuli. *J. Biomed. Opt.*, **12** (2), 021008.

71 Izzo, A.D. *et al.* (2007) Tone-on-light masking reveals cochlear spatial tuning properties of laser stimulation. Midwinter Research meeting of ARO.

72 Harris, D.M. *et al.* (2009) Optical nerve stimulation for a vestibular prosthesis. *Proc. SPIE*, **7180**, 71800R1–71800R7.

73 Rajguru S.M., Rabbitt R.D., Matic, A.I., Highstein, S.M., and Richter, C.P. (2010) Inhibitory and excitatory vestibular afferent responses induced by infrared light stimulation of hair cells. Midwinter Meeting of ARO, Anaheim, CA.

74 Teudt, I.U. *et al.* (2007) Optical stimulation of the facial nerve: a new monitoring technique? *Laryngoscope*, **117** (9), 1641–1647.

75 Jacques, S.L. (1992) Laser-tissue interactions. Photochemical, photothermal, and photomechanical. *Surg. Clin. North Am.*, **72** (3), 531–558.

76 Thomsen, S. (1991) Pathologic analysis of photothermal and photomechanical effects of laser-tissue interactions. *Photochem. Photobiol.*, **53** (6), 825–835.

77 Wells, J.D. *et al.* (2007) Biophysical mechanism responsible for low-level, transient optical stimulation of peripheral nerve. *Biophys. J.*, **92** (6). doi:10.1529/biophysj.107.104786.

78 Wells, J.D. *et al.* (2006) Biophysical mechanism responsible for pulsed low-level laser excitation of neural tissue. *Proc. SPIE*, **6084**, 60840X1–60840X7.

79 Wells, J. *et al.* (2007) Biophysical mechanisms of transient optical

stimulation of peripheral nerve. *Biophys. J.*, **93** (7), 2567–2580.

80 Caterina, M.J. *et al.* (1997) The capsaicin receptor: a heat-activated ion channel in the pain pathway. *Nature*, **389** (6653), 816–824.

81 Zheng, J. *et al.* (2003) Vanilloid receptors in hearing: altered cochlear sensitivity by vanilloids and expression of TRPV1 in the organ of corti. *J. Neurophysiol.*, **90** (1), 444–455.

82 Takumida, M. *et al.* (2005) Transient receptor potential channels in the inner ear: presence of transient receptor potential channel subfamily 1 and 4 in the guinea pig inner ear. *Acta Oto-Laryngologica*, **125**, 929–934.

83 Balaban, C.D., Zhou, J., and Li, H. (2003) Type 1 vanilloid receptor expression by mammalian inner ear ganglion cells. *Hear Res.*, **175**, 165–170.

Part Three
Tissue Optical Imaging

Laser Imaging and Manipulation in Cell Biology. Edited by Francesco S. Pavone
Copyright © 2010 WILEY-VCH Verlag GmbH & Co. KGaA, Weinheim
ISBN: 978-3-527-40929-7

7
Light–Tissue Interaction at Optical Clearing

Elina A. Genina, Alexey N. Bashkatov, Kirill V. Larin, and Valery V. Tuchin

7.1
Introduction

This chapter describes the fundamentals and advances in controlling tissue optical properties. The scattering properties of a tissue can be effectively controlled by providing matching of refractive indices of the scatterers and the ground material (i.e., optical immersion) and/or by a change of packing parameter, and/or scatterer sizing.

The reduction of light scattering by a tissue improves the image quality and precision of spectroscopic information, decreases irradiating light beam distortion, and sharpens focusing. Various physical and chemical actions, such as compression, stretching, dehydration, coagulation, and impregnation by biocompatible chemical agents are widely described in the literature as tools for controlling of tissue optical properties. As a major technology, the optical immersion method for use as exogenous optical clearing agents (OCAs) is discussed. Some important applications of the tissue immersion technique are described, such as glucose sensing, improvement of image contrast and imaging depth, laser radiation delivery, precision tissue laser photodisruption, and so on.

7.2
Light–Tissue Interaction

Light propagation within a tissue depends on the optical properties of its components: cells, cell organelles, and various fiber structures. Typically, the optical properties of tissue are characterized by the absorption and scattering coefficients, which are equal to the average number of absorption and scattering events per unit path length of a photon traveling in the tissue, and the anisotropy factor, which represents the average cosine of the scattering angles. The size, shape, and density of tissue structures, their refractive index relative to the tissue ground substance, and the polarization states of the incident light all play important roles in the propagation

Laser Imaging and Manipulation in Cell Biology. Edited by Francesco S. Pavone
Copyright © 2010 WILEY-VCH Verlag GmbH & Co. KGaA, Weinheim
ISBN: 978-3-527-40929-7

of light in tissues [1–3]. The sizes of cells and tissue structure elements vary from a few tenths of nanometers to hundreds of micrometers.

In fibrous tissues (cornea, sclera, skin dermis, dura mater, muscle, myocardium, tendon, cartilage, vessel wall, etc.) and tissues composed mostly of microfibrils and/ or microtubules, typical diameters of the cylindrical structural elements are 10–400 nm. Their length is in a range from 10–25 μm to a few millimeters [3]. The size distribution of the scattering particles may be essentially monodispersive (e.g., transparent eye cornea stroma) or quite broad, as in a turbid eye sclera [1, 2].

The collagen fibrils of fibrous tissues are arranged in individual bundles in a parallel fashion. Moreover, within each bundle the groups of fibril are separated from each other by the large empty lacunae distributed randomly in space. Collagen bundles have a wide range of widths and thicknesses. The bundles are much longer than their 0.5–8-μm diameters [4]. These ribbon-like structures are multiply cross-linked. They cross each other in all directions but remain parallel to the surface [3].

The interstitial fluid constitutes a clear, colorless liquid containing proteins, proteoglycans, glycoproteins, and hyaluronic acid. Owing to their glycosaminoglycan chains, these molecules concentrate negative charges. They are highly hydrophilic and have a propensity to attract ions, creating an osmotic imbalance that results in the glycosaminoglycan absorbing water from surrounding areas [5].

The hollow organs of the body are lined with a thin, highly cellular surface layer of epithelial tissue (oral region, maxillary sinuses, stomach, etc.), which is supported by underlying, relatively acellular connective tissue. In healthy tissue, the epithelium often consists of a single well-organized layer of cells with an *en face* diameter of 10–20 μm and height of 25 μm. In dysplastic epithelium, cells proliferate and their nuclei enlarge [3].

The outermost cellular layer of skin is epidermis, which consists of stratum corneum (SC) (mostly dead cells) and four layers of living cells. SC is a lipid-protein biphasic structure that is only 10–20 μm thick on most surfaces of the body. Owing to cell membrane keratinization, tight packing of cells, and lipid bridges between them, SC is a dense medium with poor penetration for foreign molecules [6]. Living epidermis (100 μm thick) contains most of the skin pigmentation, mainly melanin, which is produced in the melanocytes [7]. Large melanin particles, such as melanosomes (>300 nm in diameter), exhibit mainly forward scattering due to their high refractive index and relatively large size (up to 1–2 μm). In contrast, melanin dust, whose particles are small (<30 nm in diameter), provides the isotropy in the scattering profile; the optical properties of the melanin particles (30–300 nm in diameter) may be predicted by the Mie theory [8].

There are a wide variety of structures within cells that determine tissue light scattering [1–3]. Cell nuclei are on the order of 5–10 μm in diameter, mitochondria, lysosomes, and peroxisomes have dimensions of 1–2 μm, ribosomes are on the order of 20 nm in diameter, and structures within various organelles can have dimensions of up to a few hundred nanometers [3].

The absorption spectrum depends on the type of predominant absorption centers and the water content in tissues. Absorption for most tissues in the visible region is insignificant, except for the absorption bands of blood hemoglobin and some other

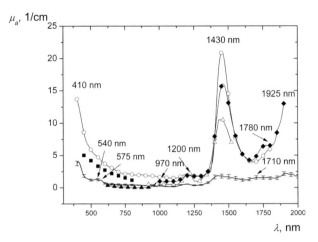

Figure 7.1 Wavelength dependence of the absorption coefficient (μ_a) of human skin *in vitro*. The solid line corresponds to the averaged experimental data, presented in Reference [9], and the vertical lines show the standard deviation values. Data taken from various sources: filled squares [10], open circles [11], filled triangles [12], open triangles [13], and filled diamonds [14].

chromophores [7]. The absorption bands of protein molecules are mainly in the near-UV region. Absorption in the IR region is essentially defined by water contained in tissues. Absolute values of absorption coefficients for typical tissues are in the range 10^{-2}–$10^4 \, \text{cm}^{-1}$ [3].

Figure 7.1 shows the wavelength dependence of the absorption coefficient of human skin presented by different authors [9–14]. The vertical lines correspond to the values of standard deviation (SD) [9].

In the figure the absorption bands of oxyhemoglobin with maxima at about 410, 540 and 575 nm are observed in the visible spectral range [9, 15]. Absorption of water in this spectral range is negligible [16]. Absorption spectra of most tissues, such as sclera, dura mater, mucosa, adipose tissues, bone, and so on, have a similar form in this spectral range [3, 9, 17–20]. In the NIR spectral range, the main chromophores are water and lipids, which are contained in different tissues in various quantities. In this spectral range, the absorption bands of water in skin with maxima at 970 [21], 1430, and 1925 nm [22] and lipids with maxima at 1710 and 1780 nm [23] are seen. At the same time, a low-intensity lipid absorption band with maximum at 930 nm [21] is not observed. The absorption band with maximum at about 1200 nm is the combination of the absorption bands of water (maximum at 1197 nm [22]) and lipids (maximum at 1212 nm [24]). The discrepancies between results obtained by different authors can be connected with the natural variability of tissue properties and the methods used for tissue preparation and storage.

Most tissues are turbid media showing a strong scattering and much less absorption (up to two orders less than scattering in the visible and NIR ranges). Moreover, tissues are rather thick. Therefore, multiple scattering is a specific feature of a wide class of tissues [1–3].

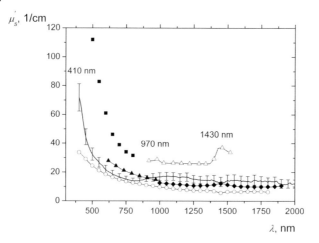

Figure 7.2 Spectral dependence of reduced scattering coefficient (μ_s') of human skin *in vitro*. The solid line corresponds to the averaged experimental data presented in Reference [9] and the vertical lines show the standard deviations. Data taken from various sources: filled squares [10], open circles [11], filled triangles [12], open triangles [13], and filled diamonds [14].

Utilizing the interaction of light with biological tissues for practical purposes depends on an understanding of the properties of two large classes of biological media. One consists of weakly scattering (transparent) tissues like cornea and the crystalline lens of eye. The other class includes strongly scattering (opaque or turbid) tissues like skin, brain, vessel wall, and eye sclera [25, 26].

Figure 7.2 presents spectral dependences of the scattering properties (shown as reduced scattering coefficient $[\mu_s' = \mu_s(1-g)]$, where μ_s is the scattering coefficient, and g is the anisotropy factor of scattering) of human skin tissue [9–14]. The anisotropy factor is usually fixed as 0.9, since this value is typical for many tissues in the visible and NIR spectral ranges [3]. The vertical lines indicate the SD values.

Figure 7.2 shows that the reduced scattering coefficient of tissues decreases with increasing wavelength in the range 400–1400 nm. However, in the spectral range from about 1400 to 2000 nm the reduced scattering coefficient can have peaks corresponding to the strong absorpation bands of water [13]. The deviation of the spectrum of the reduced scattering coefficient from a monotonic dependence can be explained by an increase of the real part of the complex refractive index of the tissue scatterers due to anomalous light dispersion. In the range of strong absorption bands, the effect produces a significant decrease of the anisotropy factor, which leads to an increase of the reduced scattering coefficient [9].

Many tissues – such as eye cornea, sclera, tendon, and cartilage, which are classified as fibrous tissues, and other structured tissues such as retina, tooth enamel, and dentin – show a wide variety of polarization properties: linear birefringence, optical activity, and linear dichroism (diattenuation). These properties are primarily defined by the tissue structure – anisotropy of form – or by the intrinsic anisotropic character of the tissue components or metabolic molecules – anisotropy of material. A large variety of tissues, such as eye cornea, tendon, cartilage, eye sclera, dura mater, testis, muscle, nerve, retina, bone, teeth, myelin, and so on, exhibit birefringence. All of these tissues contain uniaxial and/or biaxial birefringent structures [26].

Often the vector nature of light transport in scattering media, such as tissues, is ignored because of its rapid depolarization during propagation in a randomly inhomogeneous medium. However, in certain tissues (transparent eye tissues, cellular monolayers, mucous membrane, superficial skin layers, etc.), the degree of polarization of the transmitted or reflected light is measurable even when the tissue has a considerable thickness [3].

Luminescence is one of the fundamental mechanisms of interaction between light and biological objects. Luminescence is subdivided into fluorescence, corresponding to an allowed optical transition with a rather high quantum yield and a short (nanosecond) lifetime, and phosphorescence, corresponding to a "forbidden" transition with low quantum yield and long decay times in the microsecond–millisecond range [27, 28].

Fluorescence arises upon light absorption and is related to an electronic transition from the excited state to the ground state of a molecule. After excitation of biological objects by ultraviolet light ($\lambda \leq 370$ nm), fluorescence of proteins as well as of nucleic acids can be observed. Autofluorescence of proteins is related to the amino acids tryptophan, tyrosine, and phenylalanine with absorption maxima at 280, 275, and 257 nm, respectively, and emission maxima between 280 nm (phenylalanine) and 350 nm (tryptophan) [3, 27–29]. The protein spectrum is usually dominated by tryptophan. Fluorescence from collagen or elastin is excited between 400 and 600 nm with maxima around 400, 430, and 460 nm. Fluorescence of collagen and elastin can be used to distinguish various types of tissues, for example, epithelial and connective tissues [28–31].

The reduced form of coenzyme nicotinamide adenine dinucleotide (NADH) is excited selectively in the wavelength range 330–370 nm. NADH is most concentrated within mitochondria, where it is oxidized within the respiratory chain located within the inner mitochondrial membrane, and its fluorescence is an appropriate parameter for detection of ischemic or neoplastic tissues [28, 31]. Flavin mononucleotide (FMN) and dinucleotide (FAD) with excitation maxima around 380 and 450 nm have been reported to contribute to intrinsic cellular fluorescence [28].

Porphyrin molecules, for example, protoporphyrin, coproporphyrin, uroporphyrin, or hematoporphyrin, occur within the pathway of biosynthesis of hemoglobin, myoglobin, and cytochromes. Abnormalities in heme synthesis, occurring in the cases of porphyries and some hemolytic diseases, may enhance the porphyrin level within tissues considerably. Several bacteria, for example, *Propionibacterium acnes*, or bacteria within dental plaque (biofilm), such as *Porphyromonas gingivalis*, *Provotella intermedia*, and *Prevotella nigrescens*, accumulate considerable amounts of protoporphyrin [32, 33]. Therefore, acne or oral and tooth lesion detection based on measurements of intrinsic fluorescence appears to be a promising method [3].

Fluorescence spectra often give detailed information on fluorescent molecules, their conformation, binding sites, and interaction within cells and tissues. At present, various exogenous fluorescing dyes can be applied to probe cell anatomy and cell physiology [34, 35]. In humans, one such dye as indocyanine green, which is used as a diagnostic aid for blood volume determination, cardiac output, and hepatic function [36].

The depth penetration of light into a biological tissue is an important parameter for many methods of optical biomedical diagnostic and also for the correct determination of the irradiation dose in photothermal and photodynamic therapy of various diseases [1–3].

The optical transparency of tissues is maximal in the near-infrared (NIR) region, which is due to the absence, in this spectral range, of strong intrinsic chromophores that would absorb radiation in living tissues [9]. However, these tissues are characterized by rather strong scattering of NIR radiation, which prevents the attainment of clear images of localized inhomogeneities arising due to various pathologies, for example, tumor formation or the growth of microvessels.

The light penetration depth (δ) can be estimated with the relation [37]:

$$\delta = 1/\sqrt{3\mu_a\left(\mu_a + \mu_s'\right)}$$

where μ_a is the absorption coefficient and μ_s' is reduced scattering coefficient.

Table 7.1 shows the values of light penetration depth at some wavelengths used in diagnostics and photodynamic therapy for different tissues.

7.3
Tissue Clearing

In general, the scattering coefficient (μ_s) and scattering anisotropy factor (g) of a tissue depend on the refractive index mismatch between cellular tissue components: cell membrane, cytoplasm, cell nucleus, cell organelles, melanin granules, and the extracellular fluid. For fibrous (connective) tissue (eye scleral stroma, corneal stroma, skin dermis, cerebral membrane, muscle, vessel wall noncellular matrix, female breast fibrous component, cartilage, tendon, etc.) index mismatch of interstitial medium and long strands of scleroprotein (collagen-, elastin-, or reticulin-forming fibers) is important [26].

The nucleus and the cytoplasmic organelles in mammalian cells that contain similar concentrations of proteins and nucleic acids, such as the mitochondria and the ribosomes, have refractive indices that fall within a relative narrow range (1.38–1.41) [40]. The measured index for the nuclei is $n_{nc} = 1.39$ [41]. The ground matter index is usually taken as $n_0 = 1.35$–1.37. The scattering particles themselves (organelles, protein fibrils, membranes, protein globules) exhibit a higher density of proteins and lipids than the ground substance and, thus, a greater index of refraction ($n_s = 1.39$–1.47) [26]. The refractive index of the connective-tissue fibers is about 1.47, which corresponds to approximately 55% hydration of collagen, its main component, and the refractive index of the interstitial liquid is 1.35 [3].

Table 7.1 Light penetration depth (mm) for different tissues obtained at different wavelengths on the basis of data presented in References [9,17,18,20,38,39].

Tissue	Wavelengths (nm)								Reference
	532	633	675	694	780	835	930	1064	
Human skin	0.9 ± 0.1	1.7 ± 0.1	1.8 ± 0.1	1.9 ± 0.1	2.2 ± 0.1	2.3 ± 0.1	2.5 ± 0.1	3.4 ± 0.12	[9]
Human sclera	0.75 ± 0.05	1.34 ± 0.1	1.55 ± 0.15	1.64 ± 0.2	1.99 ± 0.2	2.18 ± 0.2	1.91 ± 0.1	2.56 ± 0.3	[38]
Human dura mater	1.2 ± 0.1	1.7 ± 0.1	1.7 ± 0.1	1.7 ± 0.1					[20]
Human maxillary sinuses mucous	1.3 ± 0.1	1.4 ± 0.1	1.5 ± 0.1	1.5 ± 0.1	1.6 ± 0.1	1.6 ± 0.1	1.7 ± 0.1	1.8 ± 0.1	[9]
Human stomach wall mucous	0.7 ± 0.1	1.5 ± 0.1	1.7 ± 0.1	1.7 ± 0.1	1.8 ± 0.1	1.9 ± 0.1	1.8 ± 0.1	1.8 ± 0.1	[17]
Liver		1.2 ± 0.13	1.69 ± 0.16		2.91 ± 0.3		3.68 ± 0.35		[39]
Muscle		1.47 ± 0.1	1.63 ± 0.1		3.46 ± 0.23		3.72 ± 0.29		[39]
Cranial bone	1.25 ± 0.1	1.4 ± 0.1	1.47 ± 0.1	1.5 ± 0.1	1.56 ± 0.1	1.6 ± 0.1	1.7 ± 0.1	1.8 ± 0.1	[18]

The living tissue allows one to control its optical (scattering) properties using various physical and chemical actions. Turbidity of a dispersive physical system can be effectively controlled by providing matching of refractive indices of the scatterers and the ground material. This is a so-called optical immersion technique. It is also possible to control optical properties of a disperse system by the change of its packing parameter and/or scatterer sizing.

7.3.1
Compression and Stretching

It is possible to increase significantly transmission through a soft tissue by squeezing (compressing) or stretching it [42]. The optical clarity of living tissue is due to its optical homogeneity, which is achieved through the removal of blood and interstitial liquid (water) from the compressed site. This results in a higher refractive index of the ground matter, whose value becomes close to that of scatterers (cell membrane, muscle, or collagen fibers). More close packing of tissue components at compression makes the tissue a less chaotic, more organized system, which may give less scattering due to cooperative (interference) effects. Indeed, the absence of blood in the compressed area also contributes to altered tissue absorption and refraction properties. Certain mechanisms underlying the effects of optical clearing and changing of light reflection by tissues at compression and stretching have been proposed in References [20, 43].

Several laser surgery, therapy, and diagnostic technologies include tissue compression and stretching for better transportation of the laser beam to underlying layers of tissue. The human eye compression technique allows one to perform transscleral laser coagulation of the ciliary body and retina/choroids [44].

7.3.2
Dehydration and Coagulation

Evidently, the loss of water by tissue seriously influences its optical properties. Under *in vitro* conditions spontaneous water evaporation from tissue, tissue sample heating at non-coagulating temperature, or its freezing in a refrigerator cause tissue to lose water. Typically, in the visible and NIR, far from water absorption bands, the absorption coefficient increases by a few dozen percentage points and the scattering coefficient by a few percentage points due to more close packing of tissue components caused by its shrinkage. However, the overall optical transmittance of a tissue sample increases due to a decrease of its thickness at dehydration [45]. Specifically, near the strong water absorption bands the tissue absorption coefficient decreases, due to a lower concentration of water, despite the higher density of tissue at its dehydration.

One of the major reasons for tissue dehydration *in vivo* is the action of endogenous or exogenous osmotic liquids. Dehydration induced by osmotic stimuli such as optical clearing agents (OCAs) appears to be a primary mechanism of optical clearing in collagenous and cellular tissues, whereas dehydration induces the intrinsic matching effect [46–48].

The fluid space between fibrils and organelles is filled by water and suspended salts and proteins. As water is removed from the intrafibrillar or intracellular space, soluble components of interstitial fluid become more concentrated and the refractive index increases. The resulting intrinsic refractive index matching between fibrils or organelles and their surrounding media, as well as the density of packing and particle ordering, may significantly contribute to optical clearing [46–50].

Since tissue dehydration is associated with diffusion of water from the tissue to external volume, the experimentally measured kinetics of the change of dehydration degree can be described on the basis of the equation describing diffusion through a permeable membrane. To estimate the degree of tissue dehydration under drying by hot air and/or by action of hyperosmotic agent, the following equation can be used, which describes the flow of substance from a small volume with nonzero concentration into a reservoir with zero concentration [51, 52]:

$$H_D(t) = \frac{M(t=0)-M_0}{M(t=0)}[1-\exp(-t/\tau_D)] \qquad (7.1)$$

where

M is a mass of the tissue (skin) sample (g),
M_0 corresponds to the permanent mass of the sample (i.e., dry mass) (g),
τ_D is a time constant that characterizes the rate of dehydration (s)
t is the time of dehydration (s).

The parameter M_0 includes the mass of collagen, elastin, and other components of tissue after its total dehydration, bound water, and possibility residual free water. The difference between initial and permanent mass of the sample $[M(t=0)-M_0]$ shows the amount of water escaped from tissue due to dehydration [48].

Skin dehydration kinetics induced by the stimuli of evaporation and application of OCAs is demonstrated in Figure 7.3. To enhance skin membrane permeability the SC of the samples was perforated [48, 53].

Figure 7.3 clearly shows that the degree of resulting dehydration under the influence of the osmotic agents is lower than that of air. This effect could be related to the fact that, on the one hand, glycerol, as a hyperosmotic agent, stimulates diffusion flow of free water from tissue to surrounding OCAs solution, while, on the other hand, it prevents total dehydration of tissue due to holding water inside. Glycerol is extremely hygroscopic and when left to equilibrate in a moist environment it will attract water until it reaches a level of 55% (w/w), meaning that each molecule of glycerol attracts about six molecules of water. Small hygroscopic molecules penetrate the SC where they subsequently act as a humectant. Glycols and other polyhydroxy molecules like propylene glycol, butylene glycol, glucose, saccharose, and sorbitol also work via this mechanism [54]. As glycerol viscosity decreases, it penetrates into tissue and binds molecules of water in deeper layers. In contrast, for skin dehydration in air, free water evaporates from tissue almost completely [48, 50].

Long-pulsed laser heating induces reversible and irreversible changes in the optical properties of tissue [45]. In general, the total transmittance decreases and the diffuse reflectance increases, showing nonlinear behavior during pulsed laser

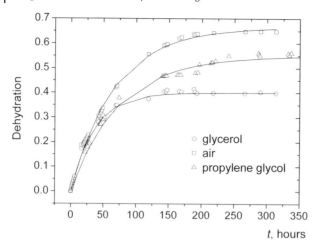

Figure 7.3 Kinetics of dehydration degree of human skin samples with damaged stratum corneum: (open squares) dehydration in air [48], (open circles) dehydration by glycerol [48], and (open triangles) dehydration by propylene glycol [53]. Symbols present experimental data, solid lines correspond to approximated dependence.

heating. Many types of tissues slowly coagulated (from 10 min to 2 h) in a hot water or saline bath (70–85 °C) exhibit an increase of their scattering and absorption coefficients [46].

7.3.3
Optical Immersion

The optical immersion technique is based on the impregnation of a tissue by a biocompatible chemical agent, which may have some hyperosmotic properties. OCAs used frequently are glucose, dextrose, fructose, glycerol, mannitol, sorbitol, propylene glycol, poly(propylene glycol), poly(ethylene glycol), 1,3-butanediol, 1,4-butanediol, and their combinations, X-ray contrasting agents (verografin, trazograph, and hypaque), and so on [20, 47–49, 52, 53, 55–63].

There are a few main mechanisms of light scattering reduction induced by an OCA [47–49, 52–64]: (i) dehydration of tissue constituents, (ii) partial replacement of the interstitial fluid by the immersion substance, and (iii) structural modification or dissociation of collagen.

The first mechanism is characteristic only for highly hyperosmotic agents. For fibrous tissue similar to sclera, dura mater, dermis, and so on, while the second mechanism is preferable for all tested chemical agents because their molecule sizes are much less than the mean cross-section of interfibrillar space, which is about 185 nm, when the diameter of the biggest molecule of PEG (20 000) is less than 5 nm [3]. Both the first and second processes mostly cause matching of the refractive indices of the tissue scatterers (cell compartments, collagen, and elastin fibers) and the cytoplasm and/or interstitial fluid.

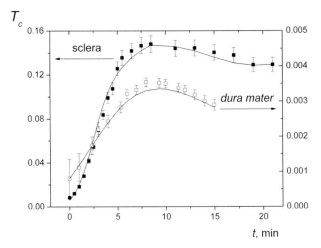

Figure 7.4 Time-dependent collimated transmittance of the human sclera and dura mater samples measured at wavelength 700nm concurrently with administration of 40%-glucose solution. Symbols correspond to the experimental data measured by a multichannel fiber-optic spectrometer LESA-6med (BioSpec, Russia). Error bars show the SD values. Solid lines correspond to the data calculated using the model presented in References [52, 63].

Figure 7.4 illustrates the kinetics of collimated transmittance of both human sclera and dura mater samples measured at wavelength 700nm concurrently with administration of aqueous 40%-glucose solution [52, 63].

It is easily seen that the untreated sclera and dura mater are poorly transparent media for visible light. Glucose administration makes these tissues highly transparent; the collimated transmittance increases 2.5–3-fold. The figure shows that the clearing process has at least two stages. At the beginning of the process the increase of the transmittance is seen, which is followed by saturation and even the decrease of the transmittance. Two major processes could take place. One of them is diffusion of glucose inside tissue and another is tissue dehydration caused by osmotic properties of the agent. In general, both processes lead to matching of refractive indices of the scatterers and the interstitial fluid that causes the decrease of tissue scattering and, therefore, the increase of the collimated transmittance. Dehydration also leads to the additional increase of optical transmission due to a decrease of tissue thickness (shrinkage) and a corresponding scatterer ordering process due to increase of particle volume fraction. However, the increase of scatterer volume fraction may also cause some competitive increase of scattering coefficient due to a random packing process (growth of particle density) that partly compensates the immersion effect. The saturation of optical clearing kinetic curves (Figure 7.4) can be explained as the saturation of glucose and water diffusion processes. Such kinetics of collimated transmission of tissue is featured not only for glucose but also for mannitol and other OCAs [52]. Some darkening at the late time period may be attributed to a few causes, such as the above-mentioned particle density growth and interaction of modified interstitial fluid (containing OCA and less water) with hydrated collagenous fibrils [46].

Structural modification also leads to tissue shrinkage, that is, to the near-order spatial correlation of scatterers and, as a result, the increased constructive interference of the elementary scattered fields in the forward direction and destructive interference in the perpendicular direction of the incident light, which may significantly increase tissue transmittance even at some refractive-index mismatch [46]. For some tissues and for the non-optimized pH of clearing agents, tissue swelling may take place that could be considered as a competitive process in providing tissue optical clearing [46, 51].

The optical clearing process in collagen-based tissues may involve a change in the supramolecular structure. Reversible solubility of collagen in sugars and sugar alcohols may take place [64]. It was demonstrated that glycerol treatment induces swelling of interfibrillar space and dissociation of collagen fibrils into microfibrils (loss of characteristic bonding in some areas). Agent-induced destabilization of collagen structures may lead to additional reduction of optical scattering in tissue due to a smaller size of the main scatterers [64].

The refractive index matching is manifested in the reduction of the scattering coefficient ($\mu_s \rightarrow 0$) and increase of single scattering directness ($g \rightarrow 1$). For fibrous tissues like skin dermis, eye sclera, dura mater, tendon, and so on, μ_s reduction can be very high [46, 49, 52, 55, 58–60]. For hematous tissue, such as the liver, its impregnation by solutes with different osmolarity also leads to refractive index matching and reduction of scattering coefficient, but the effect is not so pronounced as for fibrous tissues due to cells changing size as a result of osmotic stress [65].

For a two-component model, the mean refractive index of a tissue ($\bar{n}$) is defined by the refractive indices of its scattering centers material (n_s) and ground matter (n_0) as $\bar{n} = f_s n_s + (1 - f_s) n_0$, where f_s is the time-dependent volume fraction of the scatterers in a tissue [66]. The $n_s/n_0 \equiv m$ ratio determines the scattering coefficient. For example, the expression for the scattering coefficient, derived for a system of non-interacting thin cylinders with the number of fibrils per unit area (ϱ_s) has a form [8, 58]:

$$\mu_s \cong \varrho_s \left(\frac{\pi^5 a^4 n_0^3}{\lambda_0^3} \right) (m^2 - 1)^2 \left[1 + \frac{2}{(m^2 + 1)^2} \right] \tag{7.2}$$

where

$\varrho_s = f_{cyl}/\pi a^2$, where f_{cyl} is the surface fraction of the cylinders' faces,
a is the cylinder radius,
λ_0 is the wavelength in vacuum.

As a first approximation it is reasonable to assume that the radii of the scatterers (fibrils) and their density cannot be significantly changed by chemicals (no tissue swelling or shrinkage take place), the absolute changes of n_0 is not very high, and variations in μ_s are caused only by the change in the refractive index of the interstitial (interfibrillar) space with respect to the refractive index of the scatterers. Then, accounting for the fact that for most tissues $m \approx 1$, the ratio of the scattering coefficients at a particular wavelength as a function of refractive index ratio (m) can be written in the form [58]:

$$\mu_{s2} \cong \mu_{s1} \left(\frac{m_2 - 1}{m_1 - 1} \right)^2 \tag{7.3}$$

Indeed, this relation describes the change in tissue scattering properties due to refractive index matching or mismatching caused by changes of refractive indices of the scatterers or the background or both. Owing to square dependence, the sensitivity to indices matching is very high; for instance, only a 5% increase in the refractive index of the ground matter $(n_0 = 1.35 \rightarrow 1.42)$, when that of the scattering centers is $n_s = 1.47$, will cause a sevenfold decrease of μ_s. In the limit of equal refractive indices for non-absorbing particles and background material, $m = 1$ and $\mu_s \rightarrow 0$ [46].

Notably, for hyperosmotic agents, fluid transport within tissue is more complicated because there are at least two interacting fluxes, so the model for describing of these processes should be more complicated and should include monitoring of additional measurement parameters, such as the refractive index of the chemical agent, tissue weight and/or thickness, and osmotic pressure in a process of tissue clearing [58].

Measured values of osmotic pressure for trazograph-60 and trazograph-76 were 4.3 and 7.1 MPa, respectively. For untreated sclera, the osmotic pressure was 0.74 MPa, and increased after administration of trazograpth-60 for 30 min – up to 5.02 MPa [67]. On the one hand, the osmotic pressure causes the generation of flows and their intensities but, on the other hand, rather strong osmotic pressure may destroy tissue structure. A direct histological study has shown that there are no serious irreversible changes in the cellular and fibrous structure of the human sclera for a rather long (about 30 min) period of OCA administration [67].

Figure 7.5 presents the change in spectra of dura mater optical parameters under the action of aqueous mannitol solution ($0.16\,\mathrm{g\,ml}^{-1}$). Figures 7.5a and b show that administration of the aqueous solution of mannitol into the dura mater changes both the scattering and the absorption characteristics of this biological tissue. During clearing, the scattering decreases, on average, 1.5–2-fold in the wavelength range under study. The absorption coefficient decreases mainly in the ranges of the absorption bands of blood (Figure 7.5a). Within the Soret band, the absorption decreases, on average, 1.5-fold, and, in the range 500–600 nm, it decreases 1.2-fold [20].

This decrease in absorption is explained by a decrease in the probability of absorption of photons in an elementary volume of the tissue due to a decrease in the mean free path of photon migration within the tissue during its clearing. In addition, mannitol penetrates the vessel walls and affects the optical properties of the blood. This influence leads to a partial matching between the refractive indices of the red blood cells (erythrocytes) and the blood plasma and to a decrease in the imaginary part of the complex refractive index of the erythrocytes. This decreases both the absorption and the scattering characteristics of the blood [20].

μ_a, 1/cm

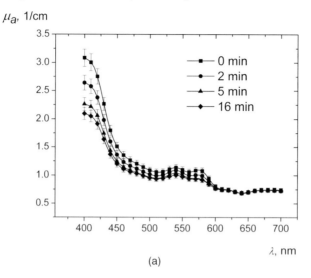

(a)

μ_s', 1/cm

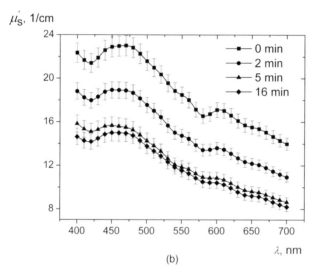

(b)

Figure 7.5 Spectra of absorption (a) and reduced scattering (b) coefficients of human dura mater sample during optical clearing under action of aqueous mannitol solution (0.16 g ml^{-1}) [20]. Bars correspond to standard deviations.

Therefore, upon penetration of OCA into the interstitial fluid of the tissue, the refractive index of the interstitial fluid increases, whereas both the real and the imaginary parts of the relative refractive indices of the scattering (collagen fibrils) and absorption (erythrocytes) centers of the tissue decrease. As a result, the scattering spectrum of the tissue in the ranges of the blood absorption bands is transformed in the course of time in such a way that the valleys become shallow and the spectral dependence of the transport scattering coefficient becomes more monotonic (Figure 7.5b) [20].

In a living tissue the relative refractive index is a function of tissue physiological or pathological state. Depending on the specificity of tissue state, the refractive index of the scatterers and/or the background may be changed (increase or decrease) and, therefore, light scattering may correspondingly increase or decrease [3, 46].

It is known that for *in vivo* application of the designed optical immersion technology additional factors such as metabolic reaction of living tissue on clearing agent application, the specificity of tissue functioning, and its physiological temperature can significantly change the kinetic characteristics and magnitude of the clearing effect [8, 49, 55, 60, 68].

Figure 7.6 presents the *in vivo* reflectance spectra of rabbit eye sclera and human skin measured at 700 nm at different time intervals after administration of a 40%-glucose solution [60, 68].

The kinetics of the polarization properties of the tissue sample at immersion can be easily observed using an optical scheme with a white light source and a tissue sample placed between two parallel or crossed polarizers. At a reduction of scattering, the degree of linearly polarized light propagating in fibrous tissue improves [26, 69].

As the tissue is immersed, the number of scattering events decreases and the residual polarization degree of transmitted linearly polarized light increases. As a result, the kinetics of the average transmittance and polarization degree of the tissue are similar. OCA-induced optical clearing leads to increasing depolarization length [69].

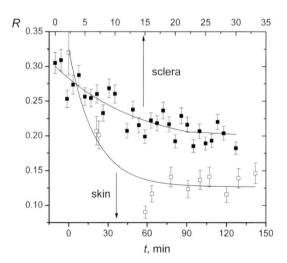

Figure 7.6 *In vivo* reflectance spectra of rabbit eye sclera and human skin measured at 700 nm at different time intervals after administration of 40%-glucose solution [60, 68]. Symbols and solid curves correspond to experimental data and their approximation, respectively.

7.4
Enhancers of Diffusion

7.4.1
Diffusion through Membranes

Several laser surgery, therapy, and noninvasive diagnostic technologies may benefit significantly from a reversible skin scattering reduction. However, slow diffusion of OCAs through a human skin barrier makes practical application of the optical immersion effect difficult.

It is well known that diffusion of aqueous solutions of substances through SC barrier is hindered. The SC is between 100 and 200 μm thick. It contains the corneocytes or horny cells, which are closely packed flat non-nucleated cells, approximately 40 μm in diameter and 0.5 μm thick. Polar structures such as corneodesmosomes contribute to SC cohesion [6]. Corneocytes are embedded in a lipid bilayer matrix. The intercellular lipids are required for a competent skin barrier and form the only continuous domain in SC [6]. Appendages such as hair follicles and sweat glands cover approximately 0.1% of the total skin surface [70].

Thus, the heterogeneous nature of SC provides some possible pathways for solute transport: appendageal, transcellular (through both corneocytes and lipid bridges), and intercellular (through the lipid phase only) [70]. Lipophilic and polar permeants are transported by the lipoidal and pore pathway of SC, respectively [71].

The pathway for water diffusion across SC has been calculated to be 50-fold longer than the thickness of SC. This suggests that permeation follows a tortuous path through the intercellular lipid matrix [6]. Penetration of compounds into the corneocytes also takes place since immersion of skin leads to swelling of the corneocytes connected with the entry of water [6]. Thus, the structure and composition of SC provides the diffusion coefficient, which is 10^3-fold less than that observed for cellular membranes [70]. The diffusion coefficient of water through SC varies in the range $2.5–8.34 \times 10^{-10} \, \mathrm{cm^2 \, s^{-1}}$ depending on the humidity of SC [71].

Diffusion of OCA through tissue layer or cell wall can be presented as diffusion of a dissolved substance through a membrane [58]. Based on Fick's first law, which limits the flux of matter (J, $\mathrm{mol \, s^{-1} \, cm^{-2}}$) to the gradient of its concentration:

$$J = -D \frac{dc}{dx} \qquad (7.4)$$

For stationary transport of matter through a thin membrane, we have [72]:

$$J = P(c_1 - c_2), \qquad (7.5)$$

where

$P = D/l$ is the coefficient of permeability, where D is the diffusion coefficient and l is the thickness of a membrane,

c_1 and c_2 are the concentrations of molecules in two spaces separated by a membrane.

Permeability of molecules across a membrane can be expressed as $P = KP_m$, where K is the partition coefficient and P_m is the permeability inside a membrane. The diffusion coefficient is a measure of the rate of entry into the cytoplasm (in the case of diffusion into a cell) or into tissue behind membrane (e.g., SC) depending on the molecular weight or size of a molecule and viscosity of a medium (e.g., the behavior of spherical molecules obeys the Stokes–Einstein law: $D = RT/6N\pi\eta r$, where N is the Avogadro constant, η is the viscosity of a medium, and r is the radius of a molecule). The term K is a measure of the solubility of the substance in lipids. A low value of K describes a molecule like water that is not soluble in lipid. This means that water, being a highly diffusive molecule, does not penetrate through lipid membranes because of its small solubility. For oils solubility is high, but diffusivity (due to high viscosity) could be low. Water and small hydrophilic molecules can penetrate only through hydrophilic pores in the membrane [72].

Owing to its fibrous structure a tissue can be presented as a porous material, which leads to modification of the chemical agent diffusion coefficient [3]:

$$D = \frac{D_i}{p} \tag{7.6}$$

Here, D_i is the chemical agent diffusion coefficient within the interstitial fluid and p is the porosity coefficient defined as:

$$p = \frac{V - V_C}{V} \tag{7.7}$$

where V is the volume of the tissue sample, and V_C is the volume of collagen fibers.

In general, biological membranes are permeable for both water and dissolved substances, but the degree of permeability for them can be quite different.

The intrinsic routes and mechanisms for transport of OCAs across the human SC and enhancement of a molecule's transport by both physical and chemical enhancers of permeation are problems under investigation.

7.4.2
Chemical Agents

As enhancers of OCA diffusivity through SC, ethanol, propylene glycol, dimethyl sulfoxide (DMSO), linoleic and oleic acids, azone, and thiazone are used [73–77].

Sometimes DMSO alone is used as the OCA due to its own extremely high permeability (as a polar aprotic solvent of SC lipids) and its high index of refraction [78].

The addition of ethanol to a solution induces a significant increase in the diffusion rate. Ethanol is a solvent known to modify the skin barrier property. The influence of ethanol on the *in vitro* transport behavior of some polar/ionic permeants in hairless mouse skin has been investigated over different ethanol/saline concentrations [73, 74]. At high concentrations (~40%) ethanol greatly enhances pore transport due to bigger pores and/or pore density of the epidermal membrane under alcohol action. This can be explained by altered or additional pore/polar pathways, which may be formed as a result of a combination of changes in the protein conformation, reorganization within the lipid polar head regions, or lipid extraction [79]. Lipid extraction may take place in conjunction with and/or independently of conformational alterations with protein domains [73].

The solvation of keratin within the SC by competition with water for the hydrogen bond binding sites and the intercalation in the polar head-groups of the lipid bilayers by propylene glycol are postulated as mechanisms for the penetration enhancing effects of propylene glycol [80]. Mixtures of different OCAs with propylene glycol increase the effectiveness of the optical clearing effect, and the clearing effect induced by propylene glycol itself is worse than that by the OCAs [78].

For thiazone and azone, a possible mechanism is an increase of fluidity of the hydrophobic SC regions and a corresponding reduction of the permeation resistance of the horny layer against drug substances [76, 77].

7.4.3
Physical Methods

Physical methods for transdermal agent delivery have two features in comparison with chemical enhancers: (i) interaction between enhancer and the agents being delivered is absent and (ii) they reduce the risk of additional skin irritation [81]. To reduce the barrier function of skin epidermis, physical methods such as tape stripping [82], microdermabrasion [83], low intensive and high intensive laser irradiation of skin surface [84, 85], iontophoresis [86], ultrasound [87] and photomechanical (shock) waves [81], needle-free injection [88], photothermal and mechanical microperforation [89, 90], or microdamaging of epidermis [91] have been proposed.

The method of tape stripping is the removal of thin strips from the skin surface using "tesa" film. The tapes are pressed onto the skin using a roller and removed with one quick movement, as described in Reference [82]. A skin strip is 3–5 μm thick.

The microdermabrasion system of aspiration–compression uses aluminium oxide crystals, which are fired from the compression system via a nozzle at a pressure of 3 bar. The hand-piece held in contact with the skin for some seconds [83].

Reference [84] utilized a Q-switched Nd:YAG laser (1064 nm) with low intensive (fluence 0.4, 0.5, and 0.6 J cm^2) irradiation together with anhydrous glycerol application to treat rat skin. The experiments showed that laser irradiation together with glycerol can significantly enhance the skin optical clearing effect. Moreover, irradiation combined with treatment of glycerol was demonstrated to be the optimum method at present.

In the method of laser ablation, a NIR laser is used in conjunction with an artificial absorber on the skin surface to create sufficient surface heating, which leads to keratinocyte disruption and skin surface ablation within the selected area-limited site [85].

Iontophoresis results in a reduction of the resistance of the skin to diffusion of ions, but as an enhancer of the flux of uncharged molecules it is a less effective method. Ultrasound and other pressure waves can affect various skin structures and inner organs due to a large penetration depth, which can be undesirable for specific tasks of OCA delivery.

A method of accelerating the penetration of OCAs due to enhancing the epidermal permeability induced by a fractional skin ablation with creation of a lattice of microzones (islets) of limited photothermal damage or lattice of islets of damage (LID) in the SC has been proposed [89]. LID are created as a result of absorption of a sufficient amount of optical energy by the lattice of microzones. The absorption leads to temperature elevation in the localized zones of interaction in contact with the skin surface. Since the ablation does not affect viable tissue, the long-term effect of fractional SC ablation is the transient deterioration of skin barrier function. This leads to the local increase of OCA penetration. The lattice of optical islets can be formed using various energy sources and delivery optics, including application of lenslet arrays, phase masks, and matrices of exogenous chromophores [48, 89].

The micro-needling method can create microchannels over a large region of skin by applying a mechanic roller. The created microchannels close within hours and the skin returns to its original condition [90]. For microdamaging of SC skin surface, rubbing with a fine 220-grit sandpaper for 2–4 min can be used [91].

For damaged SC, the dehydration of tissue accelerates (Figure 7.3) [48] and OCAs diffuse into the skin more freely; both processes lead to a refractive index matching effect.

Figure 7.7 shows a sample of human skin with a black tattoo before and after application of aqueous glycerol solution during 24 h. The SC of the sample was perforated in advance. Clearly, initially (Figure 7.7a), the tattoo was visualized significantly worse than after treatment (Figure 7.7b). The sample thickness decreased from 0.6 ± 0.05 to 0.5 ± 0.05 mm upon glycerol action [92].

Microperforation of SC, which provides effective impregnation of the upper skin layers over tattoo by an OSA, not only accelerates the skin clearing process but also promotes the increase of absorbed light fraction within the tattoo location [92, 93].

7.5
Diffusion Coefficient Estimation

Knowledge of the coefficients of diffusion and permeability is very important for the development of mathematical models describing interaction between tissues and OCAs, and for evaluation of the agent delivery rate through tissue.

Many biophysical techniques for studying the penetration of various chemicals through living tissue, and for estimation of the both diffusion and permeability coefficients, have been developed over the last 50 years. The methods are based on fluorescence measurements (including fluorescence correlation spectroscopy [94]), spectroscopic [95], Raman [96], and photoacoustic techniques [97], the usage of radioactive labels for detecting matter flux [71, 73, 74], or on the measurements of temporal changes of the scattering properties of a tissue caused by time-dependent refractive index matching [46, 52, 98], including interferometric techniques [99] and optical coherence tomography (OCT) [100, 101]. However, fluorescence techniques

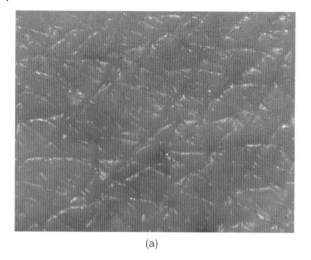

(a)

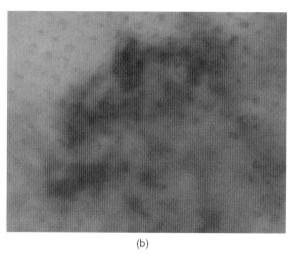

(b)

Figure 7.7 Images of skin surface with black ink tattoo; sample with tattoo before glycerol action (a) and after microperforation of skin surface and glycerol action during 24 h (b) [92].

could not be used for direct measurement of diffusion coefficients, since many of OCAs are not fluorescent. These techniques are very appropriate for measuring protein diffusivity in tissues, and since the proteins are widely used for glucose detection the techniques are then important for development of new methods, and for increasing the accuracy of existing ones, of glucose detection and monitoring. The spectroscopic, Raman, and photoacoustic methods have a great potential for measuring OCA diffusion coefficients in tissues because these methods provide excellent sensitivity for OCA detection and monitoring. Methods based on usage of radioactive labels for detecting matter flux are a "gold standard" and, thus, are widely used for measurements of chemicals and drug delivery [71, 73, 74]. However, radioactive labeling methods cannot always be used for living tissues. The measurement of the temporal changes of scattering properties of a tissue has been proposed recently for OCA diffusion and permeability coefficients estimation [52, 60, 68, 98, 100, 101].

Below, we discuss two methods based on measurement of temporal changes of the scattering properties of a tissue: (i) spectroscopy and (ii) optical coherence tomography.

7.5.1
Spectroscopic Methods

The transport of low-molecular OCAs within tissue can be described in the framework of a free diffusion model [52, 71, 73, 74, 95–98]. It is assumed that the following approximations are valid for the transport process: (i) only concentration diffusion takes place; that is, the flux of the agent into tissue at a certain point within the tissue sample is proportional to the agent concentration at this point; (ii) the diffusion coefficient is constant over the entire sample volume; (iii) penetration of the agent into a tissue sample does not change the agent concentration in the external volume; (iv) during diffusion the agent does not interact with tissue components.

Geometrically, the tissue sample can be presented as an infinite plane-parallel slab with a finite thickness. In this case, the one-dimensional diffusion problem can be considered. The one-dimensional diffusion equation of an OCA transport has a form (Fick's second law):

$$\frac{\partial C(x,t)}{\partial t} = D\frac{\partial^2 C(x,t)}{\partial x^2} \tag{7.8}$$

where

$C(x,t)$ is the agent concentration,
D is the diffusion coefficient,
t is the time,
x is the spatial coordinate.

Depending on the analytical solution used, tissue type, and the experimental setup, different kinds of initial and boundary conditions are used for studies of agent transport in tissues [52, 58, 63, 95, 98, 102].

The initial condition corresponds to the absence of an agent inside the tissue before the measurements, that is:

$$C(x,0) = 0 \tag{7.9}$$

for all inner points of the tissue sample.

When a penetrating agent is administered to tissue topically and the tissue is a semi-infinite medium, that is, $x \in [0;\infty)$, the boundary conditions have the form:

$$C(0,t) = C_0 \quad \text{and} \quad C(\infty,t) = 0 \tag{7.10}$$

Solution of Eq. (7.8) with the initial [Eq. (7.9)] and the boundary [Eq. (7.10)] conditions has the form [103]:

$$C(x,t) = C_0 \left[1 - \mathrm{erf}\left(\frac{x}{2\sqrt{Dt}} \right) \right] \tag{7.11}$$

where:

$$\mathrm{erf}(z) = \frac{2}{\sqrt{\pi}} \int_0^z \exp(-a^2)\mathrm{d}a$$

is the error function.

Experimentally, the simplest method for estimation of diffusion coefficients of liquids in tissues is based on the time-dependent measurement of collimated transmittance of tissue samples placed in an immersion liquid (Figure 7.4) [52]. Collimated transmittance could be measured in the wide wavelength range using commercially available spectrophotometers with an internal integrating sphere [46, 57–59, 61] or fiber-optic grating-array spectrometers [49, 52, 57, 60, 68, 89].

The time dependent collimated transmittance of a tissue sample impregnated by an OCA can be derived from the Bouguer–Lambert law:

$$T_c(t) = (1-R_s)^2 \exp\left\{ -[\mu_a + \mu_s(t)]l(t) \right\} \tag{7.12}$$

where

R_s is the specular reflectance,
μ_a is the tissue absorption coefficient,
$\mu_s(t)$ is the tissue time-dependent scattering coefficient,
$l(t)$ is the time-dependent thickness of the tissue sample.

The time dependence of the tissue thickness occurs due to osmotic activity of immersion agents. In general, application of OCAs can be accompanied by tissue shrinkage or swelling, which should be taken into account.

Volumetric changes of a tissue sample are mostly due to changes in sample thickness $l(t)$, which can be expressed as [52]:

$$l(t) = l(t=0) \pm A^*[1-\exp(-t/\tau_s)] \tag{7.13}$$

where $A^* = A/S$, and S is the tissue sample area. The constants A and τ_s can be obtained from direct measurements of thickness or volume of tissue samples and from time-dependent weight measurements. In the equation a "plus" sign corresponds to increasing sample volume, that is, a swelling process, and a "minus" sign corresponds to decreasing sample volume, that is, shrinkage. For example, for dura mater samples immersed in the mannitol solution, parameter A is estimated to be 0.21 and the parameter characterizing the swelling rate, that is, τ_s is 484 s [52].

By the changing volume of a tissue the swelling (shrinkage) produces the change in volume fraction of the tissue scatterers, and thus the change of the scatterer packing factor and the numerical concentration (density or volume fraction), that is, number of the scattering particles per unit area (for long cylindrical particles, density fraction) or number of the scattering particles per unit volume (for spherical particles, volume fraction). The temporal dependence of the volume fraction of the tissue scatterers is described as [52]:

$$\phi(t) = \frac{V_s}{V(t)} = \frac{\phi(t=0) \times V(t=0)}{V(t=0) \pm A[1-\exp(-t/\tau)]} \tag{7.14}$$

where V_s is the volume fraction of the tissue sample scatterers.

The optical model of many fibrous tissues can be presented as a slab with a thickness l containing scatterers (collagen fibrils) – thin dielectric cylinders with an average diameter of 10–400 nm [3], which is considerably smaller than their lengths, which can be a few millimeters. These cylinders are located in planes, which are parallel to the sample surfaces, but within each plane their orientations are random. In addition to the small so-called Rayleigh scatterers, the fibrils are arranged in individual bundles in a parallel fashion; moreover, within each bundle, the groups of fibers are separated from each other by large empty lacunae distributed randomly in space [3, 4]. In eye sclera collagen bundles show a wide range of widths (1–50 μm) and thicknesses (0.5–6 μm) [104].

For non-interacting particles the time-dependent scattering coefficient, $\mu_s(t)$, of a tissue is defined by the following equation:

$$\mu_s(t) = N(t)\sigma_s(t) \tag{7.15}$$

where $N(t)$ is the number of the scattering particles (fibrils) per unit area and $\sigma_s(t)$ is the time-dependent cross-section of scattering. The number of the scattering particles per unit area can be estimated as $N(t) = \phi(t)/(\pi a^2)$ [105], where a is the radius of the tissue scatterers. For typical intact fibrous tissues, such as sclera, dura mater, and skin dermis, ϕ is usually 0.2–0.3 [3].

To take into account interparticle correlation effects that are important for tissues with densely packed scattering particles the scattering cross-section has to be corrected by the packing factor of the scattering particles, $[1-\phi(t)]^{p+1}/[1+\phi(t)(p-1)]^{p-1}$ [106], where p is the packing dimension that describes the rate at which the empty space between scatterers diminishes as the total number density increases. For spherical particles the packing dimension is equal to 3, and the packing of sheet-like and rod-shaped particles is characterized by packing dimensions that approach 1 and 2, respectively. Thus, Eq. (7.15) has to be rewritten as:

$$\mu_s(t) = \frac{\phi(t)}{\pi a^2}\sigma_s(t)\frac{[1-\phi(t)]^3}{1+\phi(t)} \tag{7.16}$$

The time dependence of the refractive index of the interstitial fluid can be derived using the law of Gladstone and Dale, which states that the resulting value represents an average of the refractive indices of the components related to their volume fractions [66]. Such dependence is defined as:

$$n_I(t) = [1-C(t)]n_{base} + C(t)n_{oca} \tag{7.17}$$

where n_{base} is the refractive index of the tissue interstitial fluid at the initial moment, and n_{oca} is the refractive index of the OCA. The wavelength dependence of water has been presented in Reference [107] in visible and near-infrared (NIR) regions:

$$n_w(\lambda) = 1.3199 + \frac{6.878 \times 10^3}{\lambda^2} - \frac{1.132 \times 10^9}{\lambda^4} + \frac{1.11 \times 10^{14}}{\lambda^6} \tag{7.18}$$

As a first approximation, we can assume that during the interaction between the tissue and the immersion liquid the size and refractive index of the scatterers does not change. Glycosaminoglycans are charged with tissue hydration because of the formation of hydration shells. Therefore, the fibrils preferentially absorb the initial water and then remain at a relatively constant diameter [5, 51].

In this case, all changes in the tissue scattering are connected with the changes of the refractive index of the interstitial fluid described by Eq. (7.17).

The set of equations describing the dependence of OCA concentration on time represents the direct problem. Reconstruction of the diffusion coefficient of OCA can be carried out on the basis of measurement of the temporal evolution of the collimated transmittance. The solution of the inverse problem can be obtained by minimization of the target function:

$$F(D) = \sum_{i=1}^{N_t} \left[T_c(D, t_i) - T_c^*(t_i) \right]^2$$

where $T_c(D, t)$ and $T_c^*(t)$ are the calculated and experimental values of the time-dependent collimated transmittance, respectively, and N_t is the number of measurements corresponding to different time intervals at registration of the temporal kinetics of the collimated transmittance [52, 98, 108].

For *in vivo* studies, one can use the back reflectance geometry to provide measurements (Figure 7.6). For this, fiber-optic grating-array spectrometers are suitable because of their fast spectra collection during immersion agent action [29, 46, 60, 63, 68, 89]. Measured temporal evolution of the tissue optical reflectance could be used to calculate the OCA diffusion coefficient. The inverse problem can be solved by minimization of the target function:

$$F(D) = \sum_{i=1}^{N_t} \left[R(D, t_i) - R^*(t_i) \right]^2$$

where $R(D, t)$ and $R^*(t)$ are the calculated and experimental values of the time-dependent reflectance, respectively, and N_t is the number of measurements corresponding to different time intervals at registration of the temporal kinetics of the reflectance [68].

7.5.2
Optical Coherence Tomography

OCT is an imaging technique that provides images of tissues with a resolution of about 2–10 μm or less at a depth equal to or more than 1 mm depending on the optical properties of the tissue [56, 100, 101, 109–111]. It allows for determination of refractive index and attenuation coefficient in layered structures in skin and other tissues.

Attenuation of light intensity for ballistic photons, I, in a medium with scattering and absorption is described by the Bouguer–Lambert law: $I = I_0 \exp(-2\mu_t z)$, where I_0 is the incident light intensity, $\mu_t = \mu_a + \mu_s$ is the attenuation coefficient for ballistic photons, and $2z$ is the total tissue probing depth due to double pass of tissue layer by ballistic photons. Since absorption in tissues is substantially less than scattering ($\mu_a \ll \mu_s$) in the NIR spectral range, the exponential attenuation of ballistic photons in tissue depends mainly on the scattering coefficient: $I = I_0 \exp(-2\mu_s z)$ [46]. Since the scattering coefficient of tissue depends on the bulk index of refraction mismatch, an increase in refractive index of the interstitial fluid and a corresponding decrease in scattering can be detected as a change in the slope of fall-off of the depth-resolved OCT amplitude [46, 110].

The permeability coefficient for solutions in tissue layers (membranes) can be measured using two OCT methods: OCT signal slope (OCTSS) and OCT amplitude (OCTA) registration (Figure 7.8). With the OCTSS method, the average permeability coefficient of a specific region in the tissue can be calculated from the measured changes of OCT signal slope caused by agent diffusion. For this a region in the tissue, where the OCT one-dimensional distribution of intensity in-depth (A-scan) is linear and has minimal alterations, has to be selected, and its thickness (z_{region}) has to be measured. Diffusion of the agents in the chosen region can be monitored as a dynamic process (reversible) of the slope alternation, and, thus, time of diffusion can be recorded (t_{region}). The average permeability coefficient of the selected region ($\bar{P}$) can be calculated by dividing the thickness of the selected region by the time it took for the agent to diffuse through this tissue layer $\bar{P} = \left(z_{region}/t_{region}\right)$ [101, 110].

Figure 7.9 shows the OCT signal slope as a function of time recorded from sclera in a fully submersed whole eyeballs experiment during glucose diffusion [110]. Immersion of the eyeball by glucose was performed at the seventh min after the onset of the experiment. The diffusion of glucose into sclera changed the local scattering coefficient. The increase in local in-depth glucose concentration resulted in a decrease in OCT signal slope during glucose diffusion, measured from z_{region} of approximately 100 μm. As the glucose diffuses through, the reverse process starts to take place (decrease of glucose concentration as the saline diffuses in), as seen at the 45th minute of the experiment. Therefore, the calculated glucose permeability coefficient ($\bar{P}$) was approximately 5.5×10^{-6} cm s^{-1}.

The high resolution of the OCT technique allows estimation of the permeability coefficient not only as a function of time but as a function of depth as well. The OCTA method can be used to calculate the permeability coefficient at specific depths in the tissues as $P(z) = z_i/t_{z_i}$, where z_i is the depth at which measurements were performed (distance from the tissue front surface) and t_{z_i} is the time for agent diffusion to that depth. The value of t_{z_i} has to be calculated from the time point when agent was added to the tissue until the time point where agent-induced change in the OCT amplitude commenced [110]. Figure 7.10a shows a typical OCT signal measured at

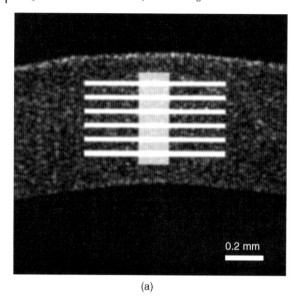

(a)

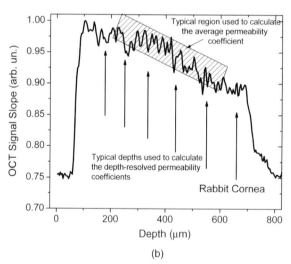

(b)

Figure 7.8 Schematic diagram showing basic principles of two methods used to calculate molecular permeability coefficients. (1) OCT signal slope (OCTSS): calculating the average permeability coefficient in tissues' stroma and computed by dividing the thickness by time – denoted by rectangles in (a) and (b); (2) OCT amplitude (OCTA): calculating the permeability coefficient at different depths, thickness from the epithelial layer to a particular depth, the time is computed from the instant glucose was added until glucose-induced change in the OCT amplitude commenced; denoted by bars in (a) and arrows in (b).

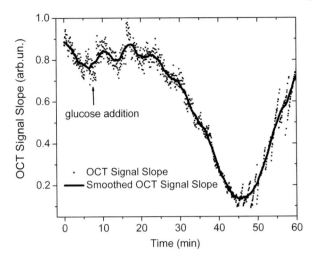

Figure 7.9 OCTSS as a function of time recorded from sclera (fully submersed whole eyeballs experiments) [110] during a glucose diffusion experiment. The arrow indicates when the agent was added.

depths of 105, 158, 225, and 273 μm from the surface of a rabbit sclera during a mannitol diffusion experiment. The arrows on each of the OCT signals depict the time the agent's action reached that particular depth, as illustrated by a sharp decrease of the OCT signal amplitude. Figure 7.10b shows the calculated permeability rates of 20%-mannitol measured at different depths in the sclera. This graph demonstrates that the permeability rate inside the sclera is not homogenous and is increasing with increasing depth, as is expected for multilayered objects with different tissue density. Therefore, the results for depth-resolved quantification of molecular diffusion in tissues demonstrate that the permeability rate inside epithelial tissues is increasing with increasing depth. Several factors are likely to contribute to such an increase in permeability rates with tissue depth: the layered structure of the tissues, the difference in diameters of the collagen fibers in each layer, and the diverse organizational patterns of the collagen bundles at different depths. For instance, the sclera possesses three distinct layers, which include the episclera, stroma, and lamina fusca. The outer layer, the episclera, consists primarily of collagen bundles that intersect at different angles along the surface of the sclera. This inconsistency in organization of the collagen bundles causes a lower permeability rate of the agent, imposing an effective resistance force that potentially reduces the speed of penetration into the tissue. In the stroma and the lamina fusca of the sclera, the collagen fibrils are more organized and oriented in two patterns, meridionally or circularly, reflecting the observed increase in agent permeability in the stroma. The different diameters of the collagen fibers at different tissue depths may also influence the diffusion processes in tissues.

Table 7.2 summarizes both the diffusivity and permeability obtained by the above-described methods.

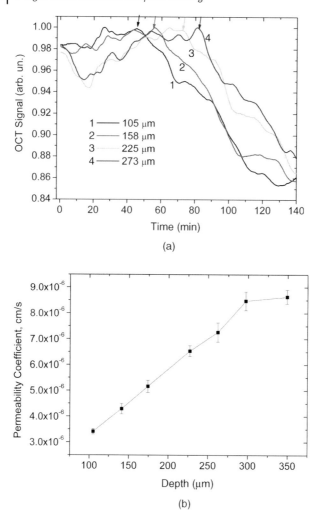

Figure 7.10 (a) OCT signal as a function of time recorded at different depths during a mannitol diffusion experiment in the sclera (arrows indicate the different depths reached by the mannitol front inside a sclera); (b) permeability rates recorded at different depths in a sclera during the diffusion of 20%-mannitol.

OCT has many advantages over other popular imaging systems such as X-ray, MRI, and ultrasound in terms of safety, cost, contrast, and resolution. However, lack of penetration depth into tissue is the main drawback of OCT. The turbidity of most biological tissues could prevent the ability of OCT to be fully engaged in various diagnostic and therapeutic procedures. As described above, application of OCAs can reduce scattering and, thus, improve imaging depth and contrast in OCT imaging of different tissues and cells. As an example, Figure 7.11 shows typical OCT images of

Table 7.2 Diffusivity (D) and permeability (P) of different OCAs in tissues obtained by different methods.

Tissue	OCA	Experimental method[a)]	$D \times 10^{-6}$ (cm^2 s^{-1})	$P \times 10^{-6}$ (cm^2 s^{-1})	Reference
Human sclera in vitro	20%-Aqueous glucose solution	CTS	0.57 ± 0.09		[98]
Human sclera in vitro	30%-Aqueous glucose solution	CTS	1.47 ± 0.36		[98]
Human sclera in vitro	40%-Aqueous glucose solution	CTS	1.52 ± 0.05		[98]
Human dura mater in vitro	20%-Aqueous glucose solution	CTS	1.63 ± 0.29		[52]
Human dura mater in vitro	20%-Aqueous mannitol solution	CTS	1.31 ± 0.41		[52]
Rat skin in vitro	40%-Aqueous glucose solution	CTS	1.1 ± 0.16		[108]
Rabbit sclera in vivo	40%-Aqueous glucose solution	ReS	0.54 ± 0.01		[68]
Human skin in vivo	40%-Aqueous glucose solution	ReS	2.56 ± 0.13		[60]
Rabbit cornea in vitro	Mannitol solution	OCT		8.99 ± 1.43	[100]
Rabbit sclera in vitro	Mannitol solution	OCT		6.18 ± 1.08	[100]
Rabbit sclera in vitro	20%-Aqueous glucose solution	OCT		8.64 ± 1.12	[100]
Pig aorta in vitro	20%-Aqueous glucose solution	OCT		14.3 ± 2.4	[110]
Pig skin in vitro	20%-Aqueous glucose solution	OCT		7.69 ± 0.56	[110]
Rhesus monkey skin in vivo	20%-Aqueous glucose solution	OCT		2.32 ± 0.2	[101]

a) Abbreviations: CTS, collimated transmittance spectroscopy; ReS, reflectance spectroscopy; OCT, optical coherence tomography.

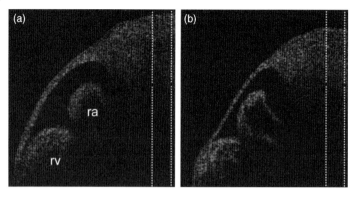

Figure 7.11 Typical OCT images of a mouse embryo acquired before (a) and 10 min after addition of 50% glycerol (b). In (a), ra = right atrium; rv = right ventricle.

the same area of a mouse embryo imaged before and 10 min after the addition of 50% glycerol (Figure 7.11a and b, respectively).

The selected imaging areas contained heart chambers as well as more optically uniform tissues used to assess the clearing effect of introduced clearing agent. As one can see from the images, during glycerol diffusion into the tissue, the atrial wall became more visually defined and the structure of the trabeculae within the ventricle became more distinguishable. Scans through the more uniform abdominal region (marked with vertical lines in Figure 7.11) were used to assess the degree of optical clearing. The percentage of signal enhancement at depths of 200 to 500 μm from the surface was calculated to be $51.5 \pm 12.5\%$ at about 10–15 min after addition of glycerol. Interestingly, the OCT signal amplitude in the upper tissue layers (up to 100 μm from the surface) is lower than that before the application of the glycerol (Figure 7.12). In the deeper layers the addition of glycerol caused an increase of the OCT signal amplitude. This result is consistent with the expectations from layer-by-layer optical clearing: the diffusion of the OCA into the tissue decreases back scattering at the upper layers by eliminating the refractive index mismatch at tissue and fluid interfaces. Consequently, because scattering is decreased in upper layers, more photons can reach deeper tissue layers, resulting in a higher OCT signal.

7.6
Applications of Tissue Optical Clearing to Different Diagnostic and Therapeutic Techniques

Over the last decade, noninvasive or minimally invasive spectroscopy and imaging techniques have found wide spread application in biomedical diagnostics, for example, OCT [3, 109–111], visible light and NIR spectroscopy [3, 112, 113], fluorescent [28–30] and polarization spectroscopy [3, 69, 114], and others. Spectroscopic

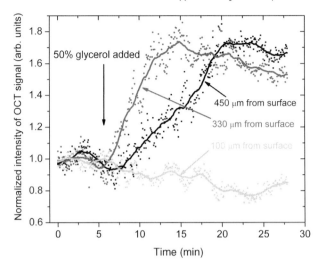

Figure 7.12 Normalized intensity of OCT signal at different depths from the surface as a function of time during addition of glycerol.

techniques can provide a deep imaging of tissues that could give information on blood oxygenation [118] and detect cutaneous and breast tumors [114, 115], whereas confocal microscopy [1, 116–118], OCT [3, 109–111], and multiphoton excitation imaging [1, 117, 119] have been used to show cellular and subcellular details of superficial living tissues. Spectroscopic, OCT, and photoacoustic techniques are applicable for blood glucose monitoring with diabetic patients [3, 120–132].

The reduction of light scattering by a tissue gives not only improvement of image quality and precision of spectroscopic information but decreases irradiating light beam distortion and gives sharp focusing.

We now discuss the application of the immersion technique to several diagnostic and therapeutic applications.

7.6.1
Glucose Sensing

7.6.1.1 NIR Technique
Measurements of blood glucose and its regulation are necessary for patients with disorders of their carbohydrate metabolism, particularly caused by diabetes mellitus [121]. In recent decades, noninvasive blood glucose monitoring has become an increasingly important topic in the realm of biomedical engineering. In particular, the introduction of optical approaches has brought exciting advances to this field [122].

NIR spectroscopy has potential in realizing noninvasive blood glucose monitoring. The NIR region is ideal for the noninvasive measurement of human body compositions because biological tissue is relatively transparent to light in this region, the so-called therapeutic window. The molecular formula of glucose is $C_6H_{12}O_6$ and several hydroxyl and methyl groups are contained in this structure. They are mainly hydrogen functional groups whose absorption occurs in the NIR region. For glucose, the second overtone absorption is in the spectral region between 1100 and 1300 nm and the first overtone absorption is in the region 1500–1800 nm [122]. In the range 1400–1500 nm there is a peak that corresponds to the absorption peak of water [133]. This information provides the theoretical basis for the measurement of blood glucose using NIR spectroscopy [122, 123]. Moreover, tissue optical clearing can be helpful for more contrast visualization of glucose bands in tissue reflectance spectra. For example, selection of an OCA without the peak in the range around 1600 nm can provide dehydration of tissue and, thereby, decrease the water peak. In contrast, glucose peak at 1600 nm can be better differentiated.

Noninvasive NIR blood glucose monitoring is a challenge because it deals with very weak signals of glucose directly from human skin, and the physiological conditions of skin that may influence on detected signal, such as body temperature, vary easily with time. However, glucose injection produces body temperature changes [134]. There is a potential for a coincidental correlation between the circadian fluctuation of glucose concentration in human blood and the circadian periodicity of the body temperature and other vital signs. Correlation between glucose concentrations and temperature-modulated localized reflectance signals at wavelengths between 590 and 935 nm, where there are no known NIR glucose absorption bands, has been presented in Reference [120].

One critical difficulty associated with *in vivo* blood glucose assay is an extremely low signal-to noise ratio of glucose peak in an NIR spectrum of human skin tissue. Therefore, the main problem of the NIR glucose monitoring is the construction of calibration models [112, 113, 122, 123].

The calibration models developed from oral glucose intake types of experiments often have chance temporal correlation. At present, the utilization of plural experiment data sets for quantitative modeling is most useful to avoid the chance temporal correlation. In Reference [112] partial least-squares regression was carried out for the NIR data (total 250 paired data-set) and calibration models were built for each subject individually. The selection of informative regions in NIR spectra for analysis can significantly refine the performance of these full-spectrum calibration techniques [113].

The floating-reference method of calibration is described in Reference [122]. The key factor and precondition of the method are the existence of the reference position where diffuse reflectance light is not sensitive to the variations of glucose concentration. Using the signal at the reference position as the internal reference for human body measurement can improve the specification for extraction of glucose information [122].

The concept of noninvasive blood glucose sensing using the scattering properties of blood is an alternative to the spectral absorption method [124]. The method of NIR frequency-domain reflectance techniques is based on changes in glucose concentration, which affects the refractive index mismatch between the interstitial fluid and tissue fibers, and hence μ_s' [125]. A glucose clamp experiment (the concentrations of injected glucose and insulin are manipulated to result in a steady concentration of glucose ever a period of time [120]) showed that $\delta\mu_s'$ at 650 nm (the distances between the source- and detector-fibers $r_{sd} = 1$–10 mm) qualitatively tracked changes in blood glucose concentration for a volunteer with diabetes [126].

The response of a non-diabetic male subject to a glucose load of 1.75 g per kg body weight, as a standard glucose tolerance test, was determined by continuously monitoring the product of $n\mu_s'$ measured on muscle tissue of the subject's thigh using a portable frequency-domain spectrometer [125]. The refractive index n of the interstitial fluid modified by glucose is defined by the equation [125] $n_{glw} = n_w + 1.515 \times 10^{-6} \times C_{gl}$, where C_{gl} is glucose concentration in mg dl^{-1} and n_w is the refractive index of water [107]. As the subject's blood glucose rose, the $n\mu_s'$ decreased. Key factors for the success of this approach are the precision of the measurements of the reduced scattering coefficient and the separation of the scattering changes from absorption changes, as obtained with the NIR frequency-domain spectrometer [125]. Evidently, other physiological effects related to glucose concentration could account for the observed variations of μ_s' and, as mentioned earlier, the effect of glucose on the blood flow in the tissue may be one of the sources of the errors of μ_s' measurements.

7.6.1.2 OCT Technique

The OCT method described in Section 7.5.2 can also be used for glucose monitoring in tissues. Since the OCT technique measures the in-depth distribution of reflected light with a high spatial resolution, changes in the in-depth distribution of the tissue scattering coefficient and/or refractive index mismatch affect the OCT signal parameters – slope and amplitude. Free blood glucose perfusion via blood capillaries and subsequent diffusion in tissues induce local changes of optical properties (scattering coefficient and refractive index); thus one can monitor and quantify the diffusion process by depth-resolved analyses of the OCT signal recorded from a tissue site [101, 110, 111, 127, 128]. Physiological glucose levels are in the range 3–30 mM, which is far below the level of making the tissue transparent. However, OCT can detect photons backscattered from different layers in the sample. Thus, changes in the scattering properties can be studied at different depths in the sample [56]. Therefore, a high sensitivity of the OCT signal from living tissues to glucose content allows one to monitor its concentration in the skin at a physiological level [127, 128].

7.6.1.3 Photoacoustic Technique

Both *in vitro* and *in vivo* studies have been carried out in the spectral range of the transparent "tissue window," around the 1–2 μm, to assess the feasibility of photoacoustic spectroscopy (PAS) for noninvasive glucose detection [129]. The photoacoustic signal is obtained by probing the sample with monochromatic radiation, which is modulated or pulsed. Absorption of probe radiation by the sample results in localized short-duration heating. Thermal expansion then gives rise to a pressure wave, which can be detected with a suitable transducer. An absorption spectrum for the sample can be obtained by recording the amplitude of generated pressure waves as a function of probe beam wavelength [129]. The pulsed PA signal is related to the properties of a turbid medium by the equation [120, 129]:

$$PA = k(\beta v^n / C_p) E_0 \mu_{eff}$$

where

PA is the signal amplitude,
k is the proportionality constant,
E_0 is the incident pulse energy,
β is the coefficient of volumetric thermal expansion,
v is the speed of sound in the medium,
C_p is the specific heat capacity,
n is a constant between one and two, depending on the particular experimental conditions,

$$\mu_{eff} = \sqrt{3\mu_a (\mu_a + \mu'_s)}.$$

In the PAS technique, the effect of glucose can be analyzed by detecting changes in the peak-to-peak value of laser-induced pressure waves [130].

The investigations have demonstrated the applicability of PAS to the measurement of glucose concentration [130, 131]. The greatest percentage change in the photo-acoustic response was observed in region of the C−H second overtone at 1126 nm, with a further peak in the region of the second O−H overtone at 939 nm [129]. In addition, the generated pulsed PA time profile can be analyzed to detect the effect of glucose on tissue scattering, which is reduced by increasing glucose concentration [46, 120].

A new concept for *in vivo* glucose detection in the mid-infrared spectral range is based on the use of two-quantum cascade lasers emitting at wavelengths 9.26 and 9.38 µm to produce PA signals in the forearm skin [132]. One of the wavelengths correlates with glucose absorption, while the other does not. Determination of glucose concentration in the extracellular fluid of the stratum spinosum permits deduction of the glucose concentration in blood, because the two factors correlate closely with each other. This method allows us to improve glucose specificity and to remove the effect of other blood substances.

7.6.1.4 Raman Spectroscopy

Raman spectroscopy (RS) can provide potentially rapid, precise, and accurate analysis of OCA concentration and biochemical composition. RS provides information about the inelastic scattering, which occurs when vibrational or rotational energy of target molecules is exchanged with incident probe radiation. Raman spectra can be utilized to identify molecules such as glucose, because these spectra are characteristic of variations in the molecular polarizability and dipole momentum. Enejder *et al.* [135] have accurately measured glucose concentrations in 17 non-diabetic volunteers following an oral glucose tolerance protocol.

In contrast to infrared and NIR spectroscopies, Raman spectroscopy has a spectral signature that is less affected by water, which is very important for tissue study. In addition, Raman spectral bands are considerably narrower (typically $10–20\,cm^{-1}$ wide [136]) than those produced in NIR spectral experiments. The RS also has the ability for simultaneous estimation of multiple analytes, requires minimum sample preparation, and would allow for direct sample analysis [137]. Like infrared absorption spectra, Raman spectra exhibit highly specific bands, which are dependent on concentration. As a rule, for Raman analysis of tissue, the spectral region between 400 and $2000\,cm^{-1}$, commonly referred to as the "fingerprint region," is employed. Different molecular vibrations lead to Raman scattering in this part of the spectrum. In many cases bands can be assigned to specific molecular vibrations or molecular species, aiding the interpretation of the spectra in terms of biochemical composition of the tissue [137].

Because of the reduction of elastic light scattering at tissue optical clearing, more effective interaction of a probing laser beam with the target molecules is expected. The OCAs increase the signal-to-noise ratio, reduce the systematic error incurred as a result of incompletely resolved surface and subsurface spectra, and significantly improve the Raman signal [96].

7.6.2
Tissue Imaging

7.6.2.1 Confocal Microscopy
Reflection confocal microscopy (RCM) is widely used at present in various biomedical investigations for visualization of the internal structure of biological tissues on a cellular and subcellular level [116, 117]. A confocal microscope illuminates and detects the scattered or fluorescent light from the same volume within the specimen. The main advantage of RCM is its ability to optically section thick specimens [117]. This technique makes it possible to obtain high-quality (with a micrometer spatial resolution) images of cellular layers. A high image contrast and a high spatial resolution of RCM are achieved due to probing a small volume of the medium bounded by the size of the central focal spot, which is formed by a focusing optical system. The main limitation of the confocal microscopy in skin studies is high scattering that distorts the quality of cell images. The increase in transparency of the upper skin layers can improve the penetration depth, image contrast, and spatial resolution of confocal microscopy [118].

Using Monte Carlo simulation of the point spread function it was shown that confocal microscopic probing of skin at optical clearing is potentially useful for deep reticular dermis monitoring and improving the image contrast and spatial resolution of the upper cell layers [118].

7.6.2.2 Nonlinear Microscopy
Optical clearing seems to be a promising technique for improvement of detected signals in nonlinear spectroscopy. In Reference [138] the authors have examined the effect of optical clearing with glycerol to achieve greater in-depth imaging of specimens of skeletal muscle tissue and mouse tendon (collagen based) by three-dimensional (3D) second-harmonic generation (SHG) imaging microscopy. It was

found that treatment with 50%-glycerol results in a 2.5-fold increase in the SHG imaging depth. In Reference [139] it was also shown that the axial attenuation of the forward SHG signal decreases with increasing glycerol concentration (25%, 50%, and 75%). This response results from the combination of the primary and secondary filter effects on the SHG creation and propagation, respectively [139]. The authors of Reference [139] note that the application of glycerol results in a swelling of the muscle cells, where the higher concentrations result in thicker tissues for the same exposure time. The tendon also exhibits significant swelling on immersion in glycerol solutions.

Thus, it was shown that reduction of the primary filter following glycerol treatment dominates the axial attenuation response in both muscle and tendon. However, these disparate tissue types have been shown to clear through different mechanisms of the glycerol–tissue interaction. In the a-cellular tendon, glycerol application reduces scattering by both index matching as well an increasing the inter-fibril separation. Through analysis of the axial response as a function of glycerol concentration in striated muscle, the authors conclude that the mechanism in this tissue arises from matching of the refractive index of the cytoplasm of the muscle cells with that of the surrounding higher index collagenous perimysium [139].

Collagen fibers produce a high second-harmonic signal and can be imaged inside skin dermis with SHG microscopy [140]. Evidently, due to optical clearing, less scattering in the epidermis for the incident long wavelength light (800 nm) and especially for the backward SHG short wavelength light (400 nm) may improve SHG images of dermis collagen structures [46].

However, at 100%-glycerol application to rodent skin dermis and tendon samples, as well as to engineered tissue model (raft), a high efficiency of tissue optical clearing was achieved in the wavelength range 400–700 nm, but the SHG signal was significantly degraded in the course of glycerol application and returned back to the initial state after tissue rehydration by application of saline [64]. The loss of SHG signal in Reference [64] is associated with the reversible dissociation of collagen fibers and the corresponding loss of fibril organization on glycerol action. Such an explanation is somewhat contradictory, because less organization of collagen fibers will lead to higher scattering [46]. Since the significant effect of optical clearing at glycerol application is tissue dehydration, an explanation following from data of Reference [140] seems to be more adequate. Using reflection-type SHG polarimetry, it was shown in Reference [140] that the SHG polarization signal for chicken skin dermis was almost unchanged and the SHG intensity was decreased to about a quarter at tissue dehydration. The authors have hypothesized that the decrease of SHG intensity results in a change of linear optical properties, that is, scattering efficiency (reduction of the SHG photon recycling), rather than of efficiency of SHG in the tissues.

It is possible that SHG and optical clearing may provide an ideal mechanism to study physiology in highly scattering skeletal or cardiac muscle tissue with significantly improved depth of penetration and achievable imaging depth. However, the problem of the change of SHG signal on the optical clearing of fibrous tissue as skin, sclera, and others requires further research.

7.6.2.3 Multiphoton Microscopy

The application of OCAs may prove to be particularly relevant for enhancing two-photon microscopy [141], since it has been shown that the effect of scattering is to drastically reduce penetration depth to less than that of the equivalent single-photon fluorescence while largely leaving resolution unchanged [117, 119]. This happens mostly due to excitation beam defocusing (distortion) in the scattering media. On the other hand, this technique is useful in understanding the molecular mechanism of tissue optical clearing upon immersion and dehydration.

The first demonstration of two-photon in-depth signal improvement using optical immersion technique with hyperosmotic agents, such as glycerol, propylene glycol, both in anhydrous form, and aqueous glucose solution, was performed by authors of Reference [141] in *ex vivo* experiments with human dermis. Such improvements were obtained within a few minutes of application. The images in Figure 7.13 show the evolution in time of a section at 60 μm depth, after immersion in glycerol [141].

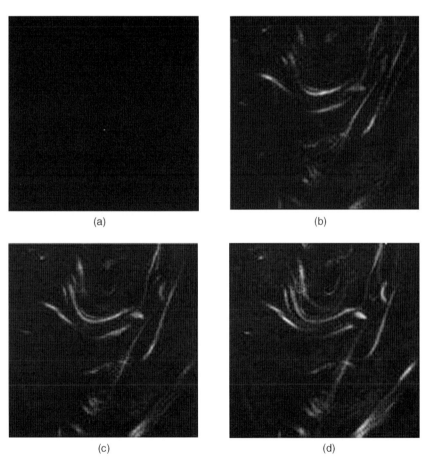

(a) (b)

(c) (d)

Figure 7.13 Time-lapse sequence of images at 60 μm depth in a skin sample, for immersion in glycerol at time $t = 0$ (a), 120 (b), 240 (c), and 420 s (d) [141].

Before data acquisition, the sample was immersed in 0.1 ml of phosphate-buffered saline (PBS) to prevent drying and shrinkage. Then, the sample was immersed in 0.5 ml of an OCA and one image stack was acquired every 30 s for 6–7 min. Finally, the OCA was removed and the sample was immersed again in 0.1 ml of PBS, to observe the reversibility of the clearing process. The upper limit of tissue shrinkage was estimated as 2% in the course of 6–7 min of OCA application. Only for glycerol was the effect shown to be partially reversible through reapplication of PBS. This adds some support to the hypothesis of alteration to the collagen structure as being a significant contributor to the clearing effect.

The average contrast in each image and relative contrast (RC) were defined as [141]:

$$\text{Contrast} = \sum_{i,j=1}^{N_{\text{lines}}} |I_{i,j} - \langle I_{i,j} \rangle|, \quad RC = 100 \frac{\text{Contrast[OCA]} - \text{Contrast[PBS]}}{\text{Contrast[PBS]}}$$

(7.19)

where $\langle I_{i,j} \rangle$ is the mean intensity of the nearest eight pixels and $N_{\text{lines}} = N - 2$, with $N = 500$; Contrast[OCA] and Contrast[PBS] are calculated using Eq. (7.19), for OCA and PBS immersion, respectively. The values of RC served for comparison purposes.

Of the three agents, glycerol was the most efficient with respect to saturation level (RC = 16.3 at 20 μm depth), but also the slowest. Propylene glycol was similarly efficient (RC = 12.6 at 20 μm depth), whereas, glucose was the worst (RC = 5.1 at 20 μm depth) but diffuses three times faster than glycerol and five times faster than propylene glycol. The RC dramatically increased with increasing depth.

The effect on deeper layers is greater because of the cumulative effect of the reduction in scattering in the superficial layers of the tissue sample, which provides less attenuation of the incident and detected fluorescent light. The contrast is also dependent on fluorescence intensity, which is proportional to the squared intensity of the excitation intensity and mostly dependent on excitation beam focusing ability. Better focusing (less focused beam distortion) is achieved in less scattering media [141].

Additional morphological information is provided by combination of SHG microscopy with two-photon excitation fluorescence microscopy [142]. Nonlinear laser imaging can be improved at tissue optical clearing but requires further investigations.

7.6.2.4 Polarized Microscopy

Practically all healthy connective and vascular tissues show the strong or weak optical anisotropy typical for either uniaxial or biaxial crystals [143]. Pathological tissues as a rule show isotropic optical properties because of irregularities of tissue structure (more chaotic cell structures and blood vessel supply network) [144]. Reduction of scattering upon optical immersion makes it possible to detect the polarization anisotropy of tissues more easily and to separate the effects of light scattering and intrinsic birefringence on the tissue polarization properties. It is also possible to study birefringence of form with optical immersion, but when the immersion is strong the average refractive index of the tissue structure is close to the index of the ground media, and the birefringence of form may be too small to be detected [26].

The OCA-induced optical clearing leads to increasing depolarization length. Clearing has a similar impact on scattering and correspondingly on the improvement of polarization properties with increasing efficiency from longer to shorter wavelengths [69]. Owing to less scattering of the longer wavelengths, the initial polarization degree is higher for these wavelengths and thus clearing efficiency is not large; however, the absolute values of achieved polarization degree of transmitted light are the highest.

With the reduction of scattering, tissue birefringence can be measured more precisely. In particular, the birefringence of form and material can be separated. For example, in a translucent human scleral sample impregnated with a highly concentrated glucose solution (about 70%), the measured optical anisotropy $\Delta n = (n_e - n_o)$ was $\sim 10^{-3}$ [26].

7.6.2.5 Optical Projection Tomography

Optical projection tomography (OPT) is a new approach for 3D imaging of small biological specimens. It fills an imaging gap between magnetic resonance imaging and confocal microscopy, being most suited to specimens that are 1–10 mm across [145].

The ability to analyze the organization of biological tissue in three dimensions has proven to be invaluable in understanding embryo development, a complex process in which tissues undergo an intricate sequence of movements relative to each other. A related goal is the mapping of gene expression patterns onto these 3D tissue descriptions. This information provides clues about the biological functions of genes and also indicates which genes may interact with each other [146]. Because OPT can record both absorption and emission profiles, it is able to image the wealth of different staining techniques that exist to record the spatial distribution of gene activity. The most common assay for gene activity at the RNA level is still a protocol that produces a purple precipitate within the tissue. Because this precipitate is not fluorescent, it cannot be imaged by confocal microscopy. However, in bright-field mode, OPT can recreate the 3D distribution of this precipitate for a complete 11.5 dpc (days-post-coitum) mouse embryo [145, 146]. Other important applications for OPT are to help analyze normal and abnormal morphology and to localize where labeled cells are within a tissue [145].

The most common approach for OPT imaging is to suspend the specimen in an index-matching liquid to reduce the scattering of light on the surface and reduce heterogeneities of refractive index throughout the specimen. This means that light passes through the specimen in approximately straight lines and a standard back-projection algorithm can generate relatively high-resolution images [145].

7.6.3
Therapeutic Applications

In vivo control of tissue optical properties can be very important for some therapeutic applications.

In Reference [68] scattering and absorption of different layers of eye under the action of 40%-glucose solution have been simulated by Monte Carlo method. The modeling has shown that the mean increase of the fraction of photons absorbed in retina at the clearing of sclera is about 30%. Thus, scleral optical clearing may provide more precise and effective coagulation of retinal pigmented epithelium and choroid layers.

Experiments have shown that OCAs (in particular, glycerol) can be used to significantly reduce the radiant exposures required for cutaneous and epidermal blood vessel coagulation [147, 148] and tattoo removal [92, 93, 149, 150].

In many cases, targeted deep blood vessels in a lesion are not sufficiently heated by incident light due to competition from absorption and scattering by other tissue constituents. The skin optical clearing method was proposed makes it possible to observe dermal blood vessel through the intact skin [147, 148]. Subcutaneous radiant exposures for vessels treated with glycerol were typically several-fold lower than untreated vessels. For example, arterioles in the 80–110-μm-diameter range in untreated skin had radiant exposure values of $\sim$12 J cm^{-2}, compared to $\sim$2 J cm^{-2} in glycerol-treated cases [147].

To estimate the effectiveness of laser radiation delivery to the area of the target absorber localization, computer simulation of the alteration of skin optical properties was carried out by Monte Carlo modeling (Figure 7.14). The optical clearing of different skin layers was simulated using Mie scattering theory [8]. For the skin image analysis, optical parameters at $\lambda = 633$ nm (He-Ne laser irradiation) were used [92]. For the target area representation, an absorbing layer in the form of cross was added to the skin model. The depth of the cross location in the model was chosen as 1 mm and its thickness was 50 μm.

The cross borders in Figure 7.14a look rather fuzzy due to high light scattering by the upper tissue layers. The optical clearing of the layers causes a significant increase in contrast of the images that improves visualization of the target area. To estimate the contrast of obtained images the following formula was used: $K = (R_1 - R_2) \times (R_1 + R_2)^{-1}$, where R_1, R_2 are the skin reflectance outside the cross area and inside it, respectively [92].

Analysis of the absorption object image shows that the largest contrast is achieved at the clearing of the upper layer ($K = 0.4$) (Figure 7.14c), which improves contrast of the image on the background of intact skin 2.6-fold ($K = 0.15$) (Figure 7.14a). Insignificant contrast increase is observed at the total immersion of skin – upper and lower layers ($K = 0.28$) (see Figure 14b), because of relatively less interaction ability of light with absorbers at less multiplicity of scattering [92].

The density of irradiation energy of a frequency-doubled Nd:YAG laser ($\lambda = 532$ nm) used for tattoo removal is 2–4 [151], 3 [150], and 2.6 J cm^{-2} [152]. For a ruby laser ($\lambda = 694$ nm) the energy density is 3.5 [153] or 4–7 J cm^{-2} [151]. The energy density of a Q-switched alexandrite laser ($\lambda = 755$ nm) is 5 [150] and 10–16 J cm^{-2} [152]. For a Q-switched Nd:YAG laser ($\lambda = 1064$ nm) the energy density is 4–8 [151] and 5 J cm^{-2} [152]. In Reference [93] it was shown that to achieve a similar result as those without skin optical clearing, when using OCA the density of laser energy can be reduced by 50–60% depending on the target area localization depth in blue-green spectral range, by 30–40% in the red spectral range, and 10–20% in the NIR spectral range. This allows a decrease in thermal damage of skin.

(a)　　　　　　　　　　(b)

(c)

Figure 7.14 Result of Monte Carlo simulation of an image of an absorption layer in skin (cross) at wavelength 633 nm; the layer is 1.0 mm deep, localization of the cross is 1.0mm in depth. Skin without clearing (a); all skin layers (excluding subcutaneous adipose layer) are immersed (b); skin layers above the cross are immersed in glycerol administered topically (c) [92].

7.7
Conclusion

This chapter has demonstrated some specific features of tissue optics and light–tissue interaction. It shows that administration of optical clearing agents allows one to control effectively optical properties of tissues. The control leads to the essential reduction of scattering and therefore causes much higher transmittance (optical clearing) and the appearance of a large amount of least-scattered and ballistic photons, allowing for successful application of different optical imaging and spectroscopic (optical biopsy) techniques for medical purposes. The kinetics of tissue optical clearing, defined, in general, by both the dehydration and agent diffusion processes, is characterized by different time responses that depend on the tissue and agents used. Swelling or shrinkage of the tissue and cells under the action of clearing agents may play an important role in the tissue clearing process.

The immersion technique has great potential for noninvasive medical diagnostics using reflectance spectroscopy, OCT, confocal, nonlinear, polarized microscopy and other methods where scattering is a serious limitation. Optical clearing can increase the effectiveness of several therapeutic and surgical methods using laser beam action on a target area hindered by depth of a tissue.

Acknowledgment

This work has been supported in part by grant 224014, PHOTONICS4LIFE of FP7-ICT-2007–2; grant 208.2008.2 of President of RF "Supporting of Leading Scientific Schools;" RF Fed. Agen. Education 2.1.1/4989, 2.2.1.1/2950, and 1.4.09; CRDF, RUB1-2932-SR-08, and RFBR N08-02-92224-NNSF (RF-P.R. China).

References

1 Tuchin, V.V. (ed.) (2002) *Handbook of Optical Biomedical Diagnostics*, SPIE Press, Bellingham, PM107.

2 Vo-Dinh, T. (ed.) (2003) *Biomedical Photonics Handbook*, CRC Press, Boca Raton.

3 Tuchin, V.V. (2007) *Tissue Optics: Light Scattering Methods and Instruments for Medical Diagnosis, SPIE Tutorial Texts in Optical Engineering*, Vol. TT38, SPIE Press, Bellingham, Washington.

4 Saidi, I.S., Jacques, S.L., and Tittel, F.K. (1995) Mie and Rayleigh modeling of visible-light scattering in neonatal skin. *Appl. Opt.*, **34** (31), 7410–7418.

5 Culav, E.M., Clark, C.H., and Merrilees, M.J. (1999) Connective tissue: matrix composition and its relevance to physical therapy. *Phys. Therapy*, **79**, 308–319.

6 Schaefer, H. and Redelmeier, T.E. (1996) *Skin Barrier: Principles of Percutaneous Absorption*, Karger, Basel.

7 Young, A.R. (1997) Chromophores in human skin. *Phys. Med. Biol.*, **42**, 789–802.

8 Bohren, C.F. and Huffman, D.R. (1983) *Absorption and Scattering of Light by Small Particles*, John Wiley & Sons, Ltd., New York.

9 Bashkatov, A.N., Genina, E.A., Kochubey, V.I., and Tuchin, V.V. (2005) Optical properties of human skin, subcutaneous and mucous tissues in the wavelength range from 400 to 2000 nm. *J. Phys. D Appl. Phys.*, **38**, 2543–2555.

10 Prahl, S.A. (1988) Light transport in tissue. Ph.D. Thesis, University of Texas at Austin.

11 Chan, E.K., Sorg, B., Protsenko, D., O'Neil, M., Motamedi, M., and Welch, A.J. (1996) Effects of compression on soft tissue optical properties. *IEEE J. Quantum Electron.*, **2**, 943–950.

12 Simpson, C.R., Kohl, M., Essenpreis, M., and Cope, M. (1998) Near-infrared optical properties of *ex vivo* human skin and subcutaneous tissues measured using the Monte Carlo inversion technique. *Phys. Med. Biol.*, **43**, 2465–2478.

13 Du, Y., Hu, X.H., Cariveau, M., Kalmus, G.W., and Lu, J.Q. (2001) Optical properties of porcine skin dermis between 900 nm and 1500 nm. *Phys. Med. Biol.*, **46**, 167–181.

14 Troy, T.L. and Thennadil, S.N. (2001) Optical properties of human skin in the near infrared wavelength range of 1000 to 2200 nm. *J. Biomed. Opt.*, **6**, 167–176.

15 Prahl, S.A. (2009) http://omlc.ogi.edu/ spectra/ accessed 1 August 2009).

16 Smith, R.C. and Baker, K.S. (1981) Optical properties of the clearest natural water (200–800 nm). *Appl. Opt.*, **20** (2), 177–184.

17 Bashkatov, A.N., Genina, E.A., Kochubey, V.I., Gavrilova, A.A., Kapralov, S.V., Grishaev, V.A., and Tuchin, V.V. (2007) Optical properties of human stomach mucosa in the spectral range from 400 to 2000 nm: prognosis for gastroenterology. *Med. Laser Appl.*, **22**, 95–104.

18 Bashkatov, A.N., Genina, E.A., Kochubey, V.I., and Tuchin, V.V. (2006) Optical properties of human cranial bone in the spectral range from 800 to 2000 nm. *Proc. SPIE*, **6163**, 616310.

19 Ugryumova, N., Matcher, S.J., and Attenburrow, D.P. (2004) Measurement of bone mineral density via light scattering. *Phys. Med. Biol.*, **49**, 469–483.

20 Genina, É.A., Bashkatov, A.N., Kochubey, V.I., and Tuchin, V.V. (2005) Optical clearing of human dura mater. *Opt. Spectrosc.*, **98** (3), 515–521.

21 McBride, T.O., Pogue, B.W., Poplack, S., Soho, S., Wells, W.A., Jiang, S., Osterberg, U.L., and Paulsen, K.D. (2002) Multispectral near-infrared tomography: a case study in compensating for water and lipid content in hemoglobin imaging of the breast. *J. Biomed. Opt.*, **7**, 72–79.

22 Palmer, K.F. and Williams, D. (1974) Optical properties of water in the near infrared. *J. Opt. Soc. Am.*, **64**, 1107–1110.

23 Martin, K.A. (1993) Direct measurement of moisture in skin by NIR spectroscopy. *J. Soc. Cosmet. Chem.*, **44**, 249–261.

24 Lauridsen, R.K., Everland, H., Nielsen, L.F., Engelsen, S.B., and Norgaard, L. (2003) Exploratory multivariate spectroscopic study on human skin. *Skin Res. Technol.*, **9**, 137–146.

25 Müller, G., Chance, B., Alfano, R., Arridge, S.R., Beuthan, J., Gratton, E., Kaschke, M.F., Masters, B.R., Svanberg, S., van der Zee, P. (eds) (1993) *Medical Optical Tomography: Functional Imaging and Monitoring*, vol. IS11, SPIE Press, Bellingham.

26 Tuchin, V.V., Wang, L.V., and Zimnyakov, D.A. (2006) *Optical Polarization in Biomedical Applications*, Springer-Verlag, New York.

27 Lakowicz, J.R. (1999) *Principles of Fluorescence Spectroscopy*, 2nd edn, Kluwer Academic/Plenum, New York.

28 Schneckenburger, H., Steiner, R., Strauss, W., Stock, K., and Sailer, R. (2002) Fluorescence technologies in biomedical diagnostics, in *Optical Biomedical Diagnostics*, vol. PM107 (ed. V.V. Tuchin), SPIE Press, Bellingham, pp. 827–874.

29 Sinichkin, Yu.P., Kollias, N., Zonios, G., Utz, S.R., and Tuchin, V.V. (2002) Reflectance and fluorescence spectroscopy of human skin *in vivo*, in *Optical Biomedical Diagnostics*, vol. PM107 (ed. V.V. Tuchin), SPIE Press, Bellingham, pp. 725–785.

30 Zhadin, N.N. and Alfano, R.R. (1998) Correction of the internal absorption effect in fluorescence emission and excitation spectra from absorbing and highly scattering media: theory and experiment. *J. Biomed. Opt.*, **3** (2), 171–186.

31 Drezek, R., Sokolov, K., Utzinger, U., Boiko, I., Malpica, A., Follen, M., and Richards-Kortum, R. (2001) Understanding the contributions of NADH and collagen to cervical tissue fluorescence spectra: modeling, measurements, and implications. *J. Biomed. Opt.*, **6** (4), 385–396.

32 Lucchina, L.C., Kollias, N., Gillies, R., Phillips, S.B., Muccini, J.A., Stiller, M.J., Trancik, R.J., and Drake, L.A. (1996) Fluorescence photography in the evaluation of acne. *J. Am. Acad. Dermatol.*, **35**, 58–63.

33 Soukos, N.S., Som, S., Abernethy, A.D., Ruggiero, K., Dunham, J., Lee, C., Doukas, A.G., and Goodson, J.M. (2005) Phototargeting oral black-pigmented bacteria. *Antimicrob. Agents Chemother.*, **49**, 1391–1396.

34 Schneckenburger, H., Gschwend, M.N., Sailer, R., Mock, H.P., and Straiss, W.S.L. (1998) Time-gated fluorescence microscopy in molecular and cellular biology. *Cell. Mol. Biol.*, **44**, 795–805.

35 Gannot, I., Garashi, A., Gannot, G., Chernomordik, V., and Gandjbakhche, A. (2003) *In vivo* quantitative three-dimensional localization of tumor labeled with exogenous specific fluorescence markers. *Appl. Opt.*, **42** (16), 3073–3080.

36 Ciamberlini, C., Guarnieri, V., Longobardi, G., Poggi, P., Donati, M.C., and Panzardi, G. (1997) Indocyanine green videoangiography using cooled CCD in central serous choroidopathy. *Biomed. Opt.*, **2**, 218–225.

37 Ritz, J.-P., Roggan, A., Isbert, C., Muller, G., Buhr, H., and Germer, C.-T. (2001) Optical properties of native and coagulated porcine liver tissue between 400 and 2400 nm. *Lasers Surg. Med.*, **29**, 205–212.

38 Bashkatov, A.N., Genina, E.A., Kochubey, V.I., Kamenskikh, T.G., and Tuchin, V.V. (2009) Optical clearing of human eye sclera. *Proc. SPIE*, **7163**, 71631.

39 Stolik, S., Delgado, J.A., Pérez, A., and Anasagasti, L. (2000) Measurement of the penetration depths of red and near infrared light in human "*ex vivo*" tissues. *J. Photochem. Photobiol. B: Biol.*, **57**, 90–93.

40 Drezek, R., Dunn, A., and Richards-Kortum, R. (1999) Light scattering from cells: finite-difference time-domain simulations and goniometric measurements. *Appl. Opt.*, **38** (16), 3651–3661.

41 Sokolov, K., Drezek, R., Gossagee, K., and Richards-Kortum, R. (1999) Reflectance spectroscopy with polarized light: is it sensitive to cellular and nuclear morphology. *Opt. Express*, **5**, 302–317.

42 Askar'yan, G.A. (1982) The increasing of laser and other radiation transport through soft turbid physical and biological media. *Sov. J. Quantum Electron.*, **9** (7), 1379–1383.

43 Guzelsu, N., Federici, J.F., Lim, H.C., Chauhdry, H.R., Ritter, A.B., and Findley, T. (2003) Measurement of skin stretch via light reflection. *J. Biomed. Opt.*, **8**, 80–86.

44 Nemati, B., Dunn, A., Welch, A.J., and Rylander, H.G. III (1998) Optical model for light distribution during transscleral cyclophotocoagulation. *Appl. Opt.*, **37** (4), 764–771.

45 Lin, W.-C., Motamedi, M., and Welch, A.J. (1996) Dynamics of tissue optics during laser heating of turbid media. *Appl. Opt.*, **35** (19), 3413–3420.

46 Tuchin, V.V. (2006) *Optical Clearing of Tissues and Blood*, SPIE Press, Bellingham, PM154.

47 Rylander, C.G., Stumpp, O.F., Milner, T.E., Kemp, N.J., Mendenhall, J.M., Diller, K.R., and Welch, A.J. (2006) Dehydration mechanism of optical clearing in tissue. *J. Biomed. Opt.*, **11** (4), 041117.

48 Genina, E.A., Bashkatov, A.N., Korobko, A.A., Zubkova, E.A., Tuchin, V.V., Yaroslavsky, I.V., and Altshuler, G.B. (2008) Optical clearing of human skin: comparative study of permeability and dehydration of intact and photothermally perforated skin. *J. Biomed. Opt.*, **13** (2), 021102.

49 Bashkatov, A.N., Korolevich, A.N., Tuchin, V.V., Sinichkin, Y.P., Genina, E.A., Stolnitz, M.M., Dubina, N.S., Vecherinski, S.I., and Belsley, M.S. (2006) *In vivo* investigation of human skin optical clearing and blood microcirculation under the action of glucose solution. *Asian J. Phys.*, **15** (1), 1–14.

50 Oliveira, L., Lage, A., Clemente, M.P., and Tuchin, V. (2009) Optical characterization and composition of abdominal wall muscle from rat. *Opt. Lasers Eng.*, **47**, 667–672.

51 Huang, Y. and Meek, K.M. (1999) Swelling studies on the cornea and sclera: the effect of pH and ionic strength. *Biophys. J.*, **77**, 1655–1665.

52 Bashkatov, A.N., Genina, E.A., Sinichkin, Yu.P., Kochubey, V.I., Lakodina, N.A., and Tuchin, V.V. (2003) Glucose and mannitol diffusion in human dura mater. *Biophys. J.*, **85** (5), 3310–3318.

53 Genina, E.A., Korobko, A.A., Bashkatov, A.N., Tuchin, V.V., Yaroslavsky, I.V., and Altshuler, G.B. (2007) Investigation of skin water loss and glycerol delivery through stratum corneum. *Proc. SPIE*, **6535**, 65351.

54 Wiechers, J.W., Dederen, J.C., and Rawlings, A.V. (2009) Moisturization mechanisms: internal occlusion by orthorhombic lipid phase stabilizers – a novel mechanism of action of skin moisturization, in *Skin Moisturization* (eds A.V. Rawlings and J.J. Leyden), Taylor & Francis, pp. 309–321.

55 Genina, E.A., Bashkatov, A.N., and Tuchin, V.V. (2009) Glucose-induced optical clearing effects in tissues and blood, in *Handbook of Optical Sensing of Glucose in Biological Fluids and Tissues* (ed. V.V. Tuchin), Taylor & Francis Group LLC, CRC Press, pp. 657–692.

56 Kinnunen, M., Myllyla, R., and Vainio, S. (2008) Detecting glucose-induced changes in in vivo and in vitro experiments with optical coherence tomography. *J. Biomed. Opt.*, **13** (2), 021111.

57 Genina, E.A., Bashkatov, A.N., and Tuchin, V.V. (2008) Optical clearing of cranial bone. *Adv. Opt. Technol.*, 267867.

58 Tuchin, V.V., Maksimova, I.L., Zimnyakov, D.A., Kon, I.L., Mavlutov, A.H., and Mishin, A.A. (1997) Light propagation in tissues with controlled optical properties. *J. Biomed. Opt.*, **2**, 401–417.

59 Vargas, G., Chan, E.K., Barton, J.K., Rylander, H.G. III, and Welch, A.J. (1999) Use of an agent to reduce scattering in skin. *Lasers Surg. Med.*, **24**, 133–141.

60 Tuchin, V.V., Bashkatov, A.N., Genina, E.A., Sinichkin, Yu.P., and Lakodina, N.A. (2001) In vivo investigation of the immersion-liquid-induced human skin clearing dynamics. *Techn. Phys. Lett.*, **27** (6), 489–490.

61 Mao, Z., Zhu, D., Hu, Y., Wen, X., and Han, Z. (2008) Influence of alcohols on the optical clearing effect of skin *in vivo*. *J. Biomed. Opt.*, **13** (2), 021104.

62 Wang, R.K. and Elder, J.B. (2002) Propylene glycol as a contrasting agent for optical coherence tomography to image gastro-intestinal tissues. *Lasers Surg. Med.*, **30**, 201–208.

63 Bashkatov, A.N., Genina, E.A., and Tuchin, V.V. (2002) Optical immersion as a tool for tissue scattering properties control, in *Perspectives in Engineering Optics* (eds K. Singh and V.K. Rastogi), Anita Publications, New Delhi, pp. 313–334.

64 Yeh, A.T., Choi, B., Nelson, J.S., and Tromberg, B.J. (2003) Reversible dissociation of collagen in tissues. *J. Invest. Dermatol.*, **121**, 1332–1335.

65 Liu, H., Beauvoit, B., Kimura, M., and Chance, B. (1996) Dependence of tissue optical properties on solute – induced changes in refractive index and osmolarity. *J. Biomed. Opt.*, **1**, 200–211.

66 Leonard, D.W. and Meek, K.M. (1997) Refractive indices of the collagen fibrils and extrafibrillar material of the corneal stroma. *Biophys. J.*, **72**, 1382–1387.

67 Kon, I.L., Bakutkin, V.V., Bogomolova, N.V., Tuchin, S.V., Zimnyakov, D.A., and Tuchin, V.V. (1997) Trazograph influence on osmotic pressure and tissue structures of human sclera. *Proc. SPIE*, **2971**, 198–206.

68 Genina, E.A., Bashkatov, A.N., Sinichkin, Yu.P., and Tuchin, V.V. (2006) Optical clearing of the eye sclera in vivo caused by glucose. *Quantum Electron.*, **36** (12), 1119–1124.

69 Zimnyakov, D.A. and Sinichkin, Yu. P. (2000) A study of polarization decay as applied to improved imaging in scattering media. *J. Opt. A: Pure Appl. Opt.*, **2**, 200–208.

70 Mollee, T.R. and Bracken, A.J. (2007) A model of solute transport through stratum corneum using solute capture and release. *Bull. Math. Biol.*, **69**, 1887–1907.

71 Blank, I.H., Moloney, J., Emslie, A.G., Simon, I., and Apt, C. (1984) The diffusion of water across the stratum corneum as a function of its water content. *J. Invest. Dermatol.*, **82**, 188–194.

72 Kotyk, A. and Janacek, K. (1977) *Membrane Transport: An Interdisciplinary Approach*, Plenum Press, New York.

73 Ghanem, A.-H., Mahmoud, H., Higuchi, W.I., Liu, P., and Good, W.R. (1992) The effects of ethanol on the transport of lipophilic and polar permeants across hairless mouse skin: methods/validation of a novel approach. *Int. J. Pharm.*, **78**, 137–156.

74 Peck, K.D., Ghanem, A.-H., and Higuchi, W.I. (1994) Hindered diffusion of polar molecules through and effective pore radii estimates of intact and ethanol treated human epidermal membrane. *Pharm. Res.*, **11**, 1306–1314.

75 Jiang, J. and Wang, R.K. (2004) Comparing the synergetic effects of oleic acid and dimethyl sulfoxide as vehicles for optical clearing of skin tissue *in vitro*. *Phys. Med. Biol.*, **49**, 5283–5294.

76 Xu, X. and Zhu, Q. (2007) Evaluation of skin optical clearing enhancement with

Azone as a penetration enhancer. *Opt. Commun.*, **279**, 223–228.

77 Zhi, Z., Han, Z., Luo, Q., and Zhu, D. (2009) Improve optical clearing of skin in vitro with propylene glycol as a penetration enhancer. *J. Innovative Opt. Health Sci.*, **2** (3), 269–278.

78 Jiang, J., Boese, M., Turner, P., and Wang, R.K. (2008) Penetration kinetics of dimethyl sulphoxide and glycerol in dynamic optical clearing of porcine skin tissue in vitro studied by Fourier transform infrared spectroscopic imaging. *J. Biomed. Opt.*, **13** (2), 021105.

79 Kurihara-Bergstrom, T., Knutson, K., De Noble, L.J., and Goates, C.Y. (1990) Percutaneous absorption enhancement of an ionic molecule by ethanol–water system in human skin. *Pharm. Res.*, **7**, 762–766.

80 Williams, A.C. and Barry, B.W. (2004) Penetration enhancers. *Adv. Drug Deliv. Rev.*, **56**, 603–618.

81 Lee, S., McAuliffe, D.J., Kollias, N., Flotte, T.J., and Doukas, A.G. (2002) Photomechanical delivery of 100-nm microspheres through the stratum corneum: implications for transdermal drug delivery. *Lasers Surg. Med.*, **31**, 207–210.

82 Weigmann, H.J., Lademann, J., Schanzer, S., Lindemann, U., Pelchrzim, R.V., Schaefer, H., and Sterry, W. (2001) Correlation of the local distribution of topically applied substances inside the stratum corneum determined by tape stripping to differences in bioavailability. *Skin Pharmacol. Appl. Skin Physiol.*, **14**, 93–103.

83 Lee, W.R., Tsai, R.Y., Fang, C.L., Liu, C.J., Hu, C.H., and Fang, J.Y. (2006) Microdermabrasion as a novel tool to enhance drug delivery via the skin: an animal study. *J. Dermatol. Surg.*, **32**, 1013–1022.

84 Liu, C., Zhi, Z., Tuchin, V.V., and Zhu, D. (2009) Combined laser and glycerol enhancing skin optical clearing. *Proc. SPIE*, **7186**, 71860.

85 Stumpp, O., Welch, A.J., and Neev, J. (2005) Enhancement of transdermal skin clearing agent delivery using a 980 nm diode laser. *Lasers Surg. Med.*, **37**, 278–285.

86 Nugroho, A.K., Li, G.L., Danhof, M., and Bouwstra, J.A. (2004) Transdermal iontophoresis of rotigotine across human stratum corneum *in vitro*: Influence of pH and NaCl concentration. *Pharm. Res.*, **21** (5), 844–850.

87 Tezel, A. and Mitragotri, S. (2003) Interaction of inertial cavitation bubbles with stratum corneum lipid bilayers during low-frequency sonophoresis. *Biophys. J.*, **85**, 3502–3512.

88 Stumpp, O. and Welch, A.J. (2003) Injection of glycerol into porcine skin for optical skin clearing with needle-free injection gun and determination of agent distribution using OCT and fluorescence microscopy. *Proc. SPIE*, **4949**, 44–50.

89 Tuchin, V.V., Altshuler, G.B., Gavrilova, A.A., Pravdin, A.B., Tabatadze, D., Childs, J., and Yaroslavsky, I.V. (2006) Optical clearing of skin using flashlamp-induced enhancement of epidermal permeability. *Lasers Surg. Med.*, **38**, 824–836.

90 Yoon, J., Son, T., Choi, E., Choi, B., Nelson, J.S., and Jung, B. (2008) Enhancement of optical skin clearing efficacy using a microneedle roller. *J. Biomed. Opt.*, **13** (2), 021103.

91 Stumpp, O., Chen, B., and Welch, A.J. (2006) Using sandpaper for noninvasive transepidermal optical skin clearing agent delivery. *J. Biomed. Opt.*, **11** (4), 041118.

92 Genina, E.A., Bashkatov, A.N., Tuchin, V.V., Altshuler, G.B., and Yaroslavski, I.V. (2008) The possibility of laser tattoo removal improvement due to skin optical clearing. *Quantum Electron.*, **38**, 580–587.

93 Bashkatov, A.N., Genina, E.A., Tuchin, V.V., and Altshuler, G.B. (2009) Skin optical clearing for improvement of laser tattoo removal. *Laser Phys.*, **19** (6), 1312–1322.

94 Weiss, M., Hashimoto, H., and Nilsson, T. (2003) Anomalous protein diffusion in living cells as seen by fluorescence correlation spectroscopy. *Biophys. J.*, **84**, 4043–4052.

95 Tsai, J.-C., Lin, C.-Y., Sheu, H.-M., Lo, Y.-L., and Huang, Y.-H. (2003) Noninvasive characterization of regional variation in drug transport into human stratum corneum in vivo. *Pharm. Res.*, **20**, 632–638.

96 Schulmerich, M.V., Cole, J.H., Dooley, K.A., Morris, M.D., Kreider, J.M., and Goldstein, S.A. (2008) Optical clearing in transcutaneous Raman spectroscopy of murine cortical bone tissue. *J. Biomed. Opt.*, **13** (2), 021108.

97 Lahjomri, F., Benamar, N., Chatri, E., and Leblanc, R.M. (2003) Study of the diffusion of some emulsions in the human skin by pulsed photoacoustic spectroscopy. *Phys. Med. Biol.*, **48**, 2729–2738.

98 Bashkatov, A.N., Genina, E.A., Sinichkin, Yu.P., Kochubei, V.I., Lakodina, N.A., and Tuchin, V.V. (2003) Estimation of the glucose diffusion coefficient in human eye sclera. *Biophysics*, **48**, 292–296.

99 Marucci, M., Ragnarsson, G., and Axelsson, A. (2006) Electronic speckle pattern interferometry: a novel non-invasive tool for studying drug transport rate and drug permeability through free films. *J. Controlled Release*, **114**, 369–380.

100 Ghosn, M.G., Tuchin, V.V., and Larin, K.V. (2007) Nondestructive quantification of analyte diffusion in cornea and sclera using optical coherence tomography. *Invest. Ophthal. Vis. Sci.*, **48**, 2726–2733.

101 Larin, K.V. and Tuchin, V.V. (2009) Monitoring of glucose diffusion in epithelial tissues with optical coherence tomography, in *Handbook of Optical Sensing of Glucose in Biological Fluids and Tissues* (ed. V.V. Tuchin), Taylor & Francis Group LLC, CRC Press, pp. 623–656.

102 Bashkatov, A.N., Genina, E.A., and Tuchin, V.V. (2009) Measurement of glucose diffusion coefficients in human tissues, in *Handbook of Optical Sensing of Glucose in Biological Fluids and Tissues* (ed. V.V. Tuchin), Taylor & Francis Group LLC, CRC Press, pp. 587–621.

103 Potts, R.O., Guzek, D.B., Harris, R.R., and McKie, J.E. (1985) A noninvasive, in vivo technique to quantitatively measure water concentration of the stratum corneum using attenuated total-reflectance infrared spectroscopy. *Arch. Dermatol. Res.*, **277**, 489–495.

104 Komai, Y. and Ushiki, T. (1991) The three-dimensional organization of collagen fibrils in the human cornea and sclera. *Invest. Ophthal. Vis. Sci.*, **32**, 2244–2258.

105 Cox, J.L., Farrell, R.A., Hart, R.W., and Langham, M.E. (1970) The transparency of the mammalian cornea. *J. Physiol.*, **210**, 601–616.

106 Schmitt, J.M. and Kumar, G. (1998) Optical scattering properties of soft tissue: a discrete particle model. *Appl. Opt.*, **37**, 2788–2797.

107 Kohl, M., Esseupreis, M., and Cope, M. (1995) The influence of glucose concentration upon the transport of light in tissue-simulating phantoms. *Phys. Med. Biol.*, **40**, 1267–1287.

108 Bashkatov, A.N., Genina, E.A., Korovina, I.V., Sinichkin, Yu. P., Novikova, O.V., and Tuchin, V.V. (2001) *In vivo* and in vitro study of control of rat skin optical properties by action of 40%-glucose solution. *Proc. SPIE*, 4241, 223–230.

109 Fujimoto, J.G. and Brezinski, M.E. (2003) Optical coherence tomography imaging, in *Biomedical Photonics Handbook* (ed. T. Vo-Dinh), CRC Press, Boca Rotan, Florida.

110 Larin, K.V. and Tuchin, V.V. (2008) Functional imaging and assessment of glucose diffusion in epithelial tissues with optical coherence tomography. *Quantum Electron.*, **6**, 551–556.

111 Larin, K.V., Ghosn, M.G., Ivers, S.N., Tellez, A., and Granada, J.F. (2007) Quantification of glucose diffusion in arterial tissues by using optical coherence tomography. *Laser Phys. Lett.*, **4**, 312–317.

112 Maruo, K., Tsurugi, M., Tamura, M., and Ozaki, Y. (2003) *In vivo* nondestructive measurement of blood glucose by near-infrared diffuse-reflectance spectroscopy. *Appl. Spectrosc.*, **57**, 1236–1244.

113 Jiang, J.H., Berry, R.J., Siesler, H.W., and Ozaki, Y. (2002) Wavelength interval selection in multicomponent spectral analysis by moving window partial least-squared regression with applications to mid-infrared and near-infrared spectroscopic data. *Anal. Chem.*, **74**, 3555–3565.

114 Yaroslavsky, A.N., Neel, V., and Anderson, R.R. (2003) Demarcation of nonmelanoma skin cancer margins in thick excisions using multispectral polarized light imaging. *J. Invest. Dermatol.*, **121**, 259–266.

115 Fantini, S., Heffer, E.L., Pera, V.E., Sassaroli, A., and Liu, N. (2005) Spatial and spectral information in optical mammography. *Technol. Cancer Res. Treatment*, **4**, 471–482.

116 Gerger, A., Koller, S., Kern, T., Massone, C., Steiger, K., Richtig, E., Kerl, H., and Smolle, J. (2005) Diagnostic applicability of in vivo confocal laser scanning microscopy in melanocytic skin tumors. *J. Invest. Dermatol.*, **124**, 493–498.

117 Masters, B.R. and So, P.T.C. (2001) Confocal microscopy and multi-photon excitation microscopy of human skin *in vivo*. *Opt. Exp.*, **8**, 2–10.

118 Dickie, R., Bachoo, R.M., Rupnick, M.A., Dallabrida, S.M., DeLoid, G.M., Lai, J., DePinho R.A., and Rogers, R.A. (2006) Three-dimensional visualization of microvessel architecture of whole-mount tissue by confocal microscopy. *Microvascullar Research*, **72** (1–2), 20–26.

119 König, K. (2000) Multiphoton microscopy in life science. *J. Microsc.*, **200**, 83–104.

120 Khalil, O.S. (2004) Non-invasive glucose measurement technologies: an update from 1999 to the dawn of the new millennium. *Diabetes Technol. Therap.*, **6**, 660–697.

121 Esenaliev, R.O. and Prough, D.S. (2009) Noninvasive monitoring of glucose concentration with optical coherence tomography, in *Handbook of Optical Sensing of Glucose in Biological Fluids and Tissues* (ed. V.V. Tuchin), Taylor & Francis Group LLC, CRC Press, pp. 563–586.

122 Heise, H.M., Lampen, P., and Marbach, R. (2009) Near-infrared reflection spectroscopy for non-invasive monitoring of glucose – established and novel strategies for multivariate calibration, in *Handbook of Optical Sensing of Glucose in Biological Fluids and Tissues* (ed. V.V. Tuchin), Taylor & Francis Group LLC, CRC Press, pp. 115–156.

123 Xu, K. and Wang, R.K. (2009) Challenges and countermeasures in NIR non-invasive blood glucose monitoring, in *Handbook of Optical Sensing of Glucose in Biological Fluids and Tissues* (ed. V.V. Tuchin), Taylor & Francis Group LLC, CRC Press, pp. 281–316.

124 Qu, J. and Wilson, B.C. (1997) Monte Carlo modeling studies of the effect of physiological factors and other analytes on the determination of glucose concentration *in vivo* by near infrared optical absorption and scattering measurements. *J. Biomed. Opt.*, **2** (3), 319–325.

125 Liu, R., Chen, W., Gu, X., Wang, R.K., and Xu, K. (2005) Chance correlation in non-invasive glucose measurement using near-infrared spectroscopy. *J. Phys. D Appl. Phys.*, **38** (15), 2675–2681.

126 Maier, J.S., Walker, S.A., Fantini, S., Franceschini, M.A., and Gratton, E. (1994) Possible correlation between blood glucose concentration and the reduced scattering coefficient of tissues in the near infrared. *Opt. Lett.*, **19**, 2062–2064.

127 Bruulsema, J.T., Hayward, J.E., Farrell, T.J., Patterson, M.S., Heinemann, L., Berger, M., Koschinsky, T., Sandahal-Christiansen, J., Orskov, H., Essenpreis, M., Schmelzeisen-Redeker, G., and Böcker, D. (1997) Correlation between blood glucose concentration in diabetics and noninvasively measured tissue optical scattering coefficient. *Opt. Lett.*, **22** (3), 190–192.

128 Larin, K.V., Eledrisi, M.S., Motamedi, M., and Esenaliev, R.O. (2002) Noninvasive blood glucose monitoring with optical coherence tomography. *Diabetes Care*, **25**, 2263–2267.

129 MacKenzie, H.A., Ashton, H.S., Spiers, S., Shen, Y., Freeborn, S.S., Hannigan, J., Lindberg, J., and Rae, P. (1999) Advances in photoacoustic noninvasive glucose testing. *Clin. Chem.*, **45**, 1587–1595.

130 Kinnunen, M. and Myllyla, R. (2008) Application of optical coherence tomography, pulsed photoacoustic technique, and time-of-flight technique to detect changes in the scattering properties of a tissue-simulating phantom. *J. Biomed. Opt.*, **13** (2), 024005.

131 Kinnunen, M. and Myllyla, R. (2005) Effect of glucose on photoacoustic signals at the wavelengths of 1064 and 532 nm in pig blood and Intralipid. *J. Physics D: Appl. Phys.*, **38**, 2654–2661.

132 Von Lilienfeld-Toal, H., Weidenmuller, M., Xhelaj, A., and Mantele, W. (2005) A novel approach to non-invasive glucose measurement by mid-infrared spectroscopy: the combination of quantum cascade lasers (QCL) and photoacoustic detection. *Vib. Spectrosc.*, **38**, 209–215.

133 Hale, G. and Querry, M.R. (1973) Optical constants of water in the 200-nm to 200-nm wavelength region. *Appl. Opt.*, **12** (3), 555–563.

134 Hillson, R.M. and Hockaday, T.D. (1982) Facial and sublingual temperature changes following intravenous glucose injection in diabetics. *Diabetes Metab.*, **8**, 15–19.

135 Enejder, A.M.K., Scecina, T.G., Oh, J., Hunter, M., Shih, W.-C., Sasic, S., Horowitz, G.L., and Feld, M.S. (2005) Raman spectroscopy for noninvasive glucose measurements. *J. Biomed. Opt.*, **10**, 031114.

136 Hanlon, E.B., Manoharan, R., Koo, T.W., Shafer, K.E., Motz, J.T., Fitzmaurice, M., Kramer, J.R., Itzkan, I., Dasari, R.R., and Feld, M.S. (2000) Prospects for in vivo Raman spectroscopy. *Phys. Med. Biol.*, **45**, R1–R59.

137 McNichols, R.J. and Coté, G.L. (2000) Optical glucose sensing in biological fluids: an overview. *J. Biomed. Opt.*, **5**, 5–16.

138 Plotnikov, S., Juneja, V., Isaacson, A.B., Mohler, W.A., and Campagnola, P.J. (2006) Optical clearing for improved contrast in second harmonic generation imaging of skeletal muscle. *Biophys. J.*, **90**, 328–339.

139 LaComb, R., Nadiarnykh, O., Carey, S., and Campagnola, P.J. (2008) Quantitative second harmonic generation imaging and modeling of the optical clearing mechanism in striated muscle and tendon. *J. Biomed. Opt.*, **13** (2), 021109.

140 Yasui, T., Tohno, Y., and Araki, T. (2004) Characterization of collagen orientation in human dermis by two-dimensional second-harmonic-generation polarimetry. *J. Biomed. Opt.*, **9** (2), 259–264.

141 Cicchi, R., Pavone, F.S., Massi, D., and Sampson, D.D. (2005) Contrast and depth enhancement in two-photon microscopy of human skin *ex vivo* by use of optical clearing agents. *Opt. Express*, **13** (7), 2337–2344.

142 Cicchi, R., Sestini, S., De Giorgi, V., Massi, D., Lotti, T., and Pavone, F.S. (2008) Nonlinear laser imaging of skin lesions. *J. BioPhotonics*, **1** (1), 62–73.

143 de Boer, J.F., Milner, T.E., and Nelson, J.S. (1999) Determination of the depth resolved Stokes parameters of light backscattered from turbid media using polarization sensitive optical coherence tomography. *Opt. Lett.*, **24**, 300–302.

144 de Boer, J.F. and Milner, T.E. (2002) Review of polarization sensitive optical coherence tomography and Stokes vector determination. *J. Biomed. Opt.*, **7** (3), 359–371.

145 Sharpe, J. (2004) Optical projection tomography. *Annu. Rev. Biomed. Eng.*, **6**, 17.1–17.20.

146 Sharpe, J., Ahlgren, U., Perry, P., Hill, B., Ross, A., Hecksher-Sørensen, J., Baldock, R., and Davidson, D. (2002) Optical projection tomography as a tool for 3D microscopy and gene expression studies. *Science*, **296**, 541–545.

147 Wang, J., Zhi, Z., Han, Z., Liu, C., Mao, Z., Wen, X., and Zhu, D. (2009) Accessing the structure and function information of deep skin blood vessels with noninvasive optical method. *Proc. SPIE*, **7176**, 71760Q–71760Q8.

148 Vargas, G., Barton, J.K., and Welch, A.J. (2008) Use of hyperosmotic chemical agent to improve the laser treatment of cutaneous vascular lesions. *J. Biomed. Opt.*, **13** (2), 021114.

149 Khan, M.H., Chess, S., Choi, B., Kelly, K.M., and Nelson, J.S. (2004) Can topically applied optical clearing agents increase the epidermal damage threshold and enhance therapeutic efficacy? *Lasers Surg. Med.*, **35**, 93–95.

150 McNichols, R.J., Fox, M.A., Gowda, A., Tuya, S., Bell, B., and Motamedi, M. (2005) Temporary dermal scatter reduction: quantitative assessment and implications for improved laser tattoo removal. *Lasers Surg. Med.*, **36**, 289–296.

151 Ross, E.V., Yashar, S., Michaud, N., Fitzpatrick, R., Geronemus, R., Tope, W.D., and Anderson, R.R. (2001) Tattoo darkening and nonresponse after laser treatment. *Arch. Dermatol.*, **137**, 33–37.

152 Prinz, B.M., Vavricka, S.R., Graf, P., Burg, G., and Dummer, R. (2004) Efficacy of laser treatment of tattoos using lasers emitting wavelengths of 532 nm, 755 nm and 1064 nm. *Br. J. Dermatol.*, **150**, 245–251.

153 Huzaira, M. and Anderson, R. (2002) Magnetite tattoos. *Lasers Surg. Med.*, **31**, 121–128.

Part Four
Laser Tissue Operation

8
Photodynamic Therapy – the Quest for Improved Dosimetry in the Management of Solid Tumors

Ann Johansson and Stefan Andersson-Engels

8.1
Introduction

Photodynamic therapy (PDT) has been investigated as a promising treatment modality for various malignant and non-malignant conditions. Already at the beginning of the twentieth century [1, 2] the PDT action was ascribed to the simultaneous presence of these three essential components: light, photosensitizer, and oxygen. Since 1978, when one of the first larger clinical studies was published on PDT for treating skin tumors [3], numerous clinical trials have shown the potential of PDT as a safe and successful treatment option (also where conventional therapies fail). Two indications that have gained great clinical success are the management of age-related macula degeneration (AMD) of the wet type [4] and non-melanoma skin malignancies (NMSM) [5, 6].

Attracting much interest is the extension of PDT to also target larger and deeper lying tissue regions has lately attracted much interest. The idea for this is that PDT would potentially offer an effective and safe treatment modality for solid tumors. PDT could be an attractive low-invasive treatment option for a large group of patients suffering from solid tumors in, for example, internal organs. Owing to the very limited penetration of light in tissue, the biggest challenge is to enable eradication of the entire tumor. This has been proven feasible by employing interstitial light delivery, also referred to as interstitial photodynamic therapy (IPDT), by utilizing several optical fibers inserted in the tumor mass; see, for example, References [7, 8]. Another important issue to consider is to minimize the treatment effects on surrounding organs at risk (OAR). Hence, to employ PDT as a tumor-selective therapy, several key issues need to be addressed: What are the relevant biological mechanisms resulting in tumor eradication? How can a PDT-dosimetry model be constructed to correctly reflect these *in vivo* treatment effects? How should the parameters of such a dosimetry model be adapted to each indication and individual treatment setting?

Extensive research during recent decades has made it clear that the effect of PDT on biological media, *in vitro* as well as *in vivo*, is a highly complex problem and influenced by many simultaneous and sometimes competing processes. In the clinic, there might be practical considerations limiting the complexity and extent of treatment

Laser Imaging and Manipulation in Cell Biology. Edited by Francesco S. Pavone
Copyright © 2010 WILEY-VCH Verlag GmbH & Co. KGaA, Weinheim
ISBN: 978-3-527-40929-7

monitoring, resulting in limited feedback data. Several PDT dosimetry models have, hence, been developed with the aim of correctly reflecting the different treatment-induced biological processes, and some of these models function very well under certain treatment conditions. In addition, PDT dosimetry models are often combined with real-time treatment monitoring, requiring techniques and instruments for measuring parameters of dosimetry relevance. In this context, we draw attention to some interesting recent reviews covering the topic of PDT from various aspects [9–12]. In this chapter, we put more emphasis on PDT dosimetry and on providing insight into some of the more established dosimetry models utilized for preclinical and clinical PDT. First, an overview of the basic mechanisms of *in vivo* PDT and of the most commonly employed photosensitizers is provided. Second, different PDT dosimetry models are discussed, followed by a presentation of clinical PDT-dosimetry. It will be seen that clinical PDT dosimetry is still relatively under-developed and that treatment protocols are employed on a more or less empirical basis. Finally, Section 8.6 is devoted to a discussion on future directions of PDT, with special emphasis on the PDT of solid tumors.

8.2
Photodynamic Reactions

PDT relies on the light-induced activation of a photosensitizer and the subsequent formation of different reactive oxygen species (ROS), which in turn cause cellular damage. In this section we describe the mechanisms of PDT-induced damage. This fundamental knowledge is obviously very important to better understand, optimize, and adjust the treatment parameters for different treatment settings and indications. Light with a wavelength tuned to match an absorption band of the photosensitizer excites it from the ground state, S_0, into a higher lying singlet state, S_1 (Figure 8.1). From here, the photosensitizer molecule either relaxes back to the ground state or crosses into a triplet state, T_1. As the transition from the triplet to the ground state is spin-forbidden, the triplet state lifetime is long, allowing the molecule to interact

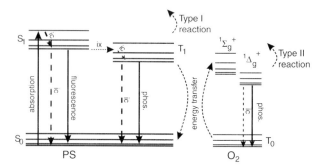

Figure 8.1 Jablonski diagram illustrating transition processes following absorption of light. PS, photosensitizer; phos, phosphorescence; ic, internal conversion; ix, inter-system crossing.

with its surroundings. The two processes that constitute the photodynamic reactions are referred to as Type I and II reactions. Type I reactions, involving electron or hydrogen atom transfer from the triplet state of the photosensitizer to substrates other than oxygen molecules, lead to the formation of highly reactive radicals or radical ions [13]. These radicals can react with oxygen to form different ROS, which oxidize tissue constituents. Type II reactions involve an electron spin exchange between T_1 and ground state oxygen molecules, $^3O_2$, leading to the formation of highly reactive singlet oxygen,$^1O_2$ [14]. As the photosensitizer returns to the ground state, one single photosensitizer molecule can generate manifold reactive species. During PDT, the photosensitizer molecules are photobleached via interactions with the ROS, thereby leading to photosensitizer degradation and the formation of various photoproducts (PPs).

The relative involvement of either Type I or II processes is influenced by factors such as the biological condition of the target, the type of photosensitizer, and its binding site within the tissue as well as the oxygen concentration at the site of activation [15, 16]. Owing to its short translational diffusion length (~ 100 nm) *in vivo* [17], the singlet oxygen distribution within the cell is essentially determined by the microlocalization of the photosensitizer. Hence, the site of photosensitizer localization also determines what cellular structures are targeted. Most cellular structures constitute potential targets for the photoinduced oxygen species. For example, in proteins certain amino acids are targeted and in lipids the unsaturated bonds are photo-oxidizable, leading to protein dysfunction, loss of enzymatic activity, and membrane damage [13]. The mechanism behind the resulting tissue damage is often categorized into three, interdependent effects: direct cell damage, vascular damage, and activation of an immune response (Figure 8.2).

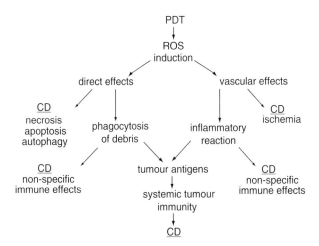

Figure 8.2 Flow chart illustrating the different mechanisms for PDT-induced tissue damage. Direct effects represent direct tumor cell damage while vascular effects denote damage to epithelial cells in the vascular system. CD: cell death.

8.2.1
Direct PDT Effects

Direct PDT effects most often consist of a combination of apoptosis, also referred to as programmed cell death, and necrosis [18]. Usually, apoptosis precedes necrosis but the exact time course, the tissue volume affected, and the relation of apoptotic versus necrotic volume depend on the photosensitizer, cell type, and light dose [18–21]. Following PDT, the zone of necrosis is sharply delineated, a fact that has led to the development of a threshold dose model (see Section 8.4). Apoptosis can often be observed beyond this necrotic boundary [19]. These peripheral regions have been exposed to lower light doses and thus it appears that the more acute the damage the more the path towards cell death is shifted in favor of necrosis [18]. Photosensitizers that bind to mitochondria are especially effective in inducing apoptosis, whereas photosensitizers that accumulate within lysosomes and plasma membranes cause less damage and usually by non-apoptotic pathways [18, 22]. Furthermore, so-called bystander effects also play a role in PDT, indicating that cells do not die independently of each other but rather that direct cell inactivation takes place in a cooperative manner [23]. Autophagy, similar to apoptosis in that it can also be considered a type of programmed cell death, has also been observed following PDT where it appears that it can act either as a survival or a death route [24]. The PDT-induced appearance of apoptosis, necrosis, bystander effects, autophagy, and other rescue pathways is controlled via a highly complex series of signaling events, either promoting or antagonizing cell death (Section 8.2.4).

8.2.2
Vascular PDT Effects

The vascular photosensitivity is related to the amount of photosensitizer present in the blood stream [25–28]. Thus, vascular-targeted PDT (VTP) relies on intra-venous (i.v.) photosensitizer injection and short times between photosensitizer and light administration, also referred to as the drug–light interval (DLI). Following light-induced oxidation, the cytoskeleton of vessel endothelial cells is rearranged, leading to exposure of the base membrane, platelet binding, and aggregation and eventually thrombosis and micro-vessel occlusion. The blood flow is thus stopped and the ensuing hypoxia and nutrient deprivation induce local tissue damage. PDT may also inhibit the production or release of nitric oxide by the endothelium, further enhancing vessel constriction [21]. However, as vascular shut-down results in hypoxia in tissue regions at larger distances from the oxygen-supplying blood vessels [29], a certain PDT-resistance might arise as no phototoxicity can be induced in the absence of oxygen.

8.2.3
Immunological Effects

Many commonly used cancer treatment modalities display pronounced immuno-suppressive effects. Although some components of the host immune system are known to accumulate photosensitizer, and hence are susceptible to PDT-induced inactivation, the PDT-induced systemic antitumor immune response is believed to be important for long-term tumor control [21, 30, 31]. PDT enables direct interaction between immune cells and tumor cells by inducing a strong inflammatory reaction and subsequent activation of tumor-specific components of the immune system. In the early inflammatory phase the treatment site is rapidly invaded by neutrophils, cytokines, mast cells, monocytes, and macrophages. The macrophages, together with antigen-presenting cells, such as dendritic cells and B cells, process and present tumor-specific antigens, a process that promotes the production of tumor-specific, cytotoxic T lymphocytes. Heat-shock proteins (HSPs), for example, HSP70, also stimulate an antitumor response. Evidence for a systemic immune response has been demonstrated in preclinical models by the observation of tumor-sensitized immune cells in lymph nodes distant from the treated lesion [32] and the observation of regressing distant, untreated metastases following PDT of the primary lesion [33]. The possibilities of manufacturing PDT-produced vaccines [34] or combining PDT with immunotherapy [30] (Section 8.2.4) to improve the overall treatment efficacy are presently being investigated.

8.2.4
Manipulating the PDT Effect

The relative importance of the mechanisms described in Sections depends on, among other factors, the rate of light delivery and the intra-tumoral localization of the photosensitizer, which is dictated by the mode of drug delivery, the DLI, the chemical character of the photosensitizer, and the tissue vascularization. Although the complex interplay is not yet fully understood there is evidence that the long-term treatment effect is dependent on all three effects [15, 35].

The fluence rate, that is, the rate of light delivery, has been shown to influence the resulting PDT effect and the mechanism by which PDT exerts its action. High fluence rates induce consumption of molecular oxygen at a rate that exceeds the rate by which oxygen can be supplied via the vasculature, thus resulting in tissue hypoxia and inhibition of further PDT action. For example, diminished tumor control has been observed for high light delivery rates in a murine colon tumor model [36]. In contrast, low fluence rates conserve oxygen levels, leading to improved PDT efficacy. Metronomic PDT (mPDT) constitutes the extreme case, where light and photosensitizer are delivered at very low rates over extended time periods (∼days) [37]. The influence of the fluence rate on the PDT effect is further discussed in Section 8.4.

Vascular-targeted PDT, as exploited by the use of short DLI, has been shown to induce pronounced vascular damage and good long-term tumor regression [28]. In contrast, a longer DLI allows the photosensitizer to diffuse further from the blood vessels. Together with the characteristic leakiness of the tumor vasculature, the so-called enhanced penetration and retention effect (EPR), see also Section 8.3.2, a longer DLI is often associated with increased tumor-selectivity. The relative contribution of the vascular effects to the total PDT result can be manipulated by carefully choosing an appropriate DLI. The use of fractionated photosensitizer administration, for example, one dose at a long DLI and the second shortly before commencing therapeutic irradiation, might offer a possibility to optimize the PDT effect by exploiting both the pronounced vascular damage and the increased photosensitizer selectivity [38]. In such a treatment situation, it might be necessary to employ initially low fluence rates to maintain vascular function and oxygen supply while targeting tissue regions distant to blood vessels. During the second phase of the treatment, higher fluence rates and light doses might then be applied to achieve vascular shutdown typical of VTP.

Extensive research has investigated the potential of manipulating cellular signaling events by suppressing various rescue pathways and/or by means of photoimmunotherapy (PIT) where different components of the PDT-induced immune response are promoted via the administration of immunoadjuvants [18, 20, 22, 39]. An enhanced expression of angiogenic and survival signals, for example, vascular endothelial growth factor (VEGF), matrix metalloproteinases (MMPs), and cyclooxygenase-2 (COX-2), has been observed post-PDT [39, 40], likely due to treatment-induced hypoxia, inflammation, and vessel destruction. Hence, the PDT itself might lead to the formation of new vessels, increased tumor invasion, and the occurrence of metastases [41]. However, the combination of PDT and anti-angiogenic and/or anti-inflammatory drugs, for example, inhibitors of VEGF and COX-2, has been shown to increase the PDT effectiveness without increasing normal tissue photosensitization in animal tumor models [40–42]. Clinically, PDT has been combined with Avastin or Lucentis, known VEGF inhibitors, and vatalanib, a tyrosine kinase inhibitor with specificity for VEGF, for the treatment of AMD, resulting in pronounced synergistic effects [43, 44].

Different approaches towards PIT include PDT combined with microbial adjuvants, cytokines, or adjuvants that augment the cellular arm of the antitumor response [30, 45]. Hence, one can influence the level of macrophages and regulatory T cells, the degree of neutrophil activation, and the amount of vessel destruction. Clinically, PIT has been employed for the treatment of malignant and non-malignant skin diseases, such as actinic keratosis (AK), superficial basal cell carcinoma (BCC), squamous cell carcinoma (SCC), and malignant melanoma [44], for high-grade vulval intraepithelial neoplasia (VIN) [46] and for genital Bowenoid papulosis (BP) [47]. For these indications, PDT has been combined with topical application of imiquimod [44], a substance that has FDA approval for the treatment of external genital and perianal warts. Imiquimod acts by stimulating both the innate immune response and the cellular arm of acquired immunity and also exhibits some anti-angiogenic characteristics. Finally, synergistic effects have also been observed for the combination of ALA-mediated PDT and Bacille Calmette-Guérin (BCG) administration for bladder cancer treatment [48].

Apoptosis seems to play a large role for the resulting PDT effect, at least within preclinical models. Hence, the effects of inhibiting various anti-apoptotic signaling events in connection to PDT are being investigated in cell cultures and animal tumor models. Substances known to influence the occurrence of apoptosis are, for example, members of the Bcl-2 family, certain stress proteins, NF$\varkappa$B, cell surface receptors involved in the apoptotic pathway, such as Fas and tumor necrosis factor (TNF), mitogen-activated protein kinase (MAPK), certain caspases, and nitric oxygen. However, translational studies evaluating modulators for apoptosis are still relatively sparse. Of potential clinical relevance might be the increased apoptotic response observed in an animal model following PDT after the concurrent administration of ursodeoxycholic acid (UDCA) [49], a substance used clinically for the treatment of biliary cirrhosis and other conditions in gastroenterology. The synergistic effect has been attributed to an increased photosensitivity of Bcl-2 in the presence of UDCA but its clinical relevance remains to be elucidated. In addition, one publication reports on the development of an immunohistochemical assay for caspase-3 activation following Pc4-mediated PDT of cutaneous T-cell lymphoma [50]. Preliminary clinical results suggest that the amount of PDT-induced apoptosis may correlate with treatment effect. The intriguing observation that the overall PDT effect is not impaired by blocking the apoptotic pathway might motivate the use of PDT for the treatment of tumors that are resistant to certain apoptosis-inducing agents, as sometimes observed, for example, for chemo- and radiation therapy [22]. Initial clinical trials investigating the possible synergistic effects of PDT combined with chemotherapy, also referred to as photochemotherapy, have been reported for cardiac, skin, and bladder cancer [39].

Several basic questions, such as the optimal timing between PDT and administration of the immunoadjuvant, whether necrosis or apoptosis is more effective in stimulating an immune response, and the clinically achievable effectiveness of PIT, still need to be answered. The potential for increased treatment efficacy provided by modulating PDT-induced processes might be good but one should also acknowledge the difficulties of transferring existing knowledge from cell cultures or animal tumor models to the clinical setting.

8.3
Photosensitizers

The term photosensitization means, as indicated above, that biological tissue is made sensitive to light by the administration of a substance. This substance may undergo various chemical reactions causing biological effects following light excitation. Much of the work on photosensitizers originates from the work on porphyrins and their relatives. However, other and quite different molecular structures have been investigated and evaluated in the search for an ideal photosensitizer. More recently, other substances, such as nanoparticles, liposome-encapsulated photosensitizers, and photoimmunoconjugates have been studied, to develop photosensitizers with improved photophysical and pharmacokinetic properties.

As there are many requirements for an ideal photosensitizer in order to result in local destruction of the target tissue, no substance has satisfactorily met all of them. In addition, different clinical settings may weigh the importance of the photosensitizer properties differently. These are the reasons why so many agents are being

Table 8.1 Clinically approved PDT and some selected early-phase clinical PDT studies.

Trade name	Photosensitizer	Excitation wavelength (nm)	Route of administration	Indication
Photofrin, Photobarr	Porfimer sodium	630	i.v.	Approved for palliative treatment of bladder, esophageal and lung cancer as well as for Barrett's oesophagus in North America and Europe
Visodyne	Bensoporphyrin (BPD) (Verteporfin)	690	i.v.	Approved for AMD in over 60 countries worldwide
Foscan	Mesotetrahydroxy phenyl chlorine (mTHPC) (Temoporfin)	652	i.v.	Approved for palliative treatment of head and neck SCC in Europe, investigated for prostate cancer
Metvix	Methyl aminolevulinate (MAL)	635	Topical	Approved for dermatological applications in approx. 30 countries worldwide, including most European countries, Australia, and New Zealand
Levulan Kerastick	5-Aminolevulinic acid (ALA)	635	Topical	Approved for AK of the face and scalp in the USA
LS11 Laserphyrin	Mono-L-aspartyl chlorine e6 (NPe6)	664	i.v.	Approved for early stage bronchopulmonary lung cancer in Japan
LuTex	Motexafin lutetium	734	i.v.	Phase I studies for prostate cancer
Pc-4	Tetrasulfonated phthalocyanine	670	i.v.	Potentially for cutaneous or subcutaneous metastasis
WST11 Padeliporfin	Palladium bacteriopheophorbide	755	i.v.	In clinical trials for AMD and prostate cancer

developed, evaluated, and employed. So far the clinically most successful PDT applications have been for the management of AMD and NMSM. Other indications for which PDT has obtained clinical approval are shown in Table 8.1 and are discussed in Reference [15]. This table also lists some photosensitizers currently being investigated for PDT of certain malignancies. Clearly, a wide span of different photophysical and pharmacokinetic properties are required for PDT of these indications.

The most serious shortcoming from a clinical point of view might be the suboptimal pharmacokinetics. Most clinically employed photosensitizers lack a truly tumor-selective uptake and are associated with prolonged, generalized photosensitivity. The latter effect is particularly pronounced for the so-called first-generation photosensitizers, for example, HpD, Photofrin, where the patient might remain photosensitive for months following systemic exposure to the compound. Some second-generation photosensitizers, for example, bensoporphyrin (BPD) and palladium bacteriopheophorbide as employed for VTP or 5-aminolaevulinic acid (ALA)-induced protoporphyrin IX (PpIX), are characterized by significantly faster clearance.

The basic properties that characterize an ideal photosensitizing compound are the absorption wavelength (important as it determines the light penetration in tissue, see below), quantum efficiency, selective accumulation in the diseased tissue, a fast clearance, and a low toxicity. The quantum efficiency is, apart from the photophysical properties, also dependent on the aggregation state of the dye, its micro-localization when activated and on the localized concentration of molecules quenching the generated excited state in the photosensitizer. The fast clearance rate should ideally include all exposed normal tissues, and also the liver, kidney, and spleen, but the clearance from the diseased tissue should preferably be slow. The toxicity of a photosensitizer is partly connected to the selectivity of the agent, meaning that photosensitizers that very selectively accumulate mainly in a small tissue volume lead to less side effects on normal tissues.

In this section, we first discuss some of the essential photophysical properties for a photosensitizer for PDT, followed by the equally important pharmacokinetic properties determining the localization of the drug within the tissue. As will be seen, many properties might be influenced by varying the basic molecular structure, by introducing or altering side-chains, by introducing various metal ions within the molecule, and by employing different delivery vehicles. We will also look at the different routes of administration and side effects. These properties will mainly be discussed in the context of photosensitizers used in clinical or preclinical studies. At the end of the section a brief introduction to some ideas for novel PDT agents will be given.

8.3.1
Photophysical Properties

The photophysical properties of a photosensitizer determine its production efficiency of ROS following light activation (Figure 8.1). A high extinction coefficient, ε, at the therapeutic wavelength leads to efficient excitation of the photosensitizer, meaning that lower light and drug doses are required, provided that the other properties are kept constant.

Absorption bands at long wavelengths provide the potential for increased treatment depth as compared to absorption at shorter wavelengths due to the lower light attenuation of biological media in the red and NIR wavelength regions. In contrast, absorption at shorter wavelengths might be useful to limit treatment depths as desirable for superficial malignancies in, for example, the esophagus. For Type II reactions, the therapeutic wavelength should correspond to an energy well exceeding that of the singlet-oxygen state to allow an efficient energy transfer from the photosensitizer to a nearby oxygen molecule, thus requiring excitation wavelengths below approximately 850 nm. By decreasing the number of double bonds in the planar-aromatic molecules or by introducing metalation, the absorption bands of some photosensitizer classes can be redshifted and/or strengthened [51].

Furthermore, metalation influences the conversion efficiencies between the photosensitizer excited states, $S_1 \rightarrow T_1$, the lifetime of the triplet state, τ_t, and the yield of singlet oxygen, Φ_Δ. For example, adding aluminium or another diamagnetic cation improves the triplet state yield and τ_t of phthalocyanines, whereas the compound ZnPc has essentially zero yield of singlet oxygen [51].

Despite the fact that photosensitizer fluorescence slightly decreases the triplet state yield, fluorescence constitutes an attractive feature as it can be utilized as a tool for tissue diagnosis, the study of photosensitizer pharmacokinetics, and PDT dosimetry (Section 8.4).

8.3.2
Pharmacokinetics and Tumor Selectivity

Photosensitizer administration paths are i.v. injection, oral intake, or topical application. One has to be aware that all delivery paths involve systemic photosensitization to some degree, although the effect is minimized for the topical administration path. Here the total dose can be kept quite low, while the local concentration can be sufficiently high.

When supplied systemically to the blood, photosensitizers accumulate in organs rich in reticuloendothelial components, that is, phagocytic cells. The highest photosensitizer levels are thus found in the liver, spleen, and kidneys followed by the lungs and heart, skin and muscle, and finally brain tissue [52].

Topical applications are most commonly used for ALA-induced PpIX. The zwitterionic character of the ALA molecule makes it highly water soluble but also limits its skin permeability. Hence, different ALA-ester derivatives with enhanced lipophilic characters have been developed to improve biological availability and skin permeability. One of these, methyl amino-leavulinate (m-ALA) [53, 54], is marketed as Metvix® and Levulan® Kerastick in Europe and in the USA, respectively. ALA- and m-ALA-PDT constitutes a good treatment option for superficial, that is, less than 2-mm thick, BCC (80–97% cure rates), AK (69–100% cure rates), and Bowen's disease (75–100% cure rates) [55, 56]. Improved treatment outcomes have also been attempted via polymeric and liposome-based drug delivery systems [54]. However, *in vivo* it has been difficult to provide evidence for improved depth penetration and to reproduce the promising results obtained *in vitro* with ALA-esters and other delivery vehicles [54]. In addition, for bladder cancer diagnostics and therapy, topical administration is employed by instilling the organ with ALA or hexyl-ALA (h-ALA, Hexvix) solutions. There is also a thermogel formulation available for mTHPC [57].

So far the only PDT drug used with oral administration is ALA as employed for diagnostic procedures (for instance for guiding the surgeon during brain tumor resections), and not so extensively for treatment purposes. Once taken up by the mucosa, the drug will be distributed in the body via the blood supply.

The lipophilicity of the photosensitizer greatly affects its pharmacokinetics. Lipophilic molecules tend to aggregate, that is, form dimeric or oligomeric compounds, in aqueous surroundings. Apart from altering the absorption and fluorescent properties and decreasing the triplet state lifetime and the yield of singlet oxygen, thus decreasing the PDT efficiency, aggregation influences the distribution within biological media. Lipophilic photosensitizers localize intracellularly, thus reaching biological structures that constitute effective targets for the photodynamic action [35]. In contrast, hydrophilic photosensitizers mostly accumulate in the vascular stroma and hence cause more vascular damage than do lipophilic photosensitizers [58]. Amphiphilic photosensitizers, that is, photosensitizers that possess both hydrophilic and lipophilic properties, have been shown to be more photodynamically active than photosensitizers exhibiting either hydrophilic or lipophilic characteristics [59].

High concentration and selectivity are thus the goals. Regarding photosensitizer distribution, one strives for both high absolute concentration and high tumor selectivity. The photosensitizer levels in surrounding normal tissue and OAR should be low. Although largely dependent on photosensitizer and tissue type, photosensitizer retention and treatment-induced tissue damage are far from tumor selective. The sometimes observed selectivity is most likely not due to any particular property of the tumor cells as compared to normal cells but rather due to certain physiological prerequisites that promote photosensitizer accumulation within tumor tissue. For example, tumors are often characterized by decreased pH, a condition that renders a photosensitizer more water soluble and increases retention. Small molecular sizes might benefit from the leaky vasculature and poor lymphatic drainage, that is, the EPR effect, in tumor tissue. Furthermore, tumor-associated macrophages can accumulate large amounts of aggregated photosensitizer. Increased selectivity can also be achieved for lipophilic photosensitizers as they more readily bind to low-density lipoproteins (LDLs) and thus make use of the increased number of LDL receptors presented in tumors [60]. By pre-associating the photosensitizer with LDL [60] or incorporating it into liposomes [61], which interact with serum lipoproteins, an improved tumor-selective uptake has been observed.

Another parameter that might influence the tumor selectivity is the DLI (Section 8.2.4). More recently investigated strategies for improving tumor-selective localization are based on active targeting, realized, for example, by linking either the photosensitizer [62] or the liposome encapsulating the photosensitizer [61] with antigen-specific monoclonal antibodies. These approaches will be discussed further in Section 8.6.

8.4
PDT Dosimetry Models

The use of PDT dosimetry is being proposed as a means to improve treatment outcome by fully employing the knowledge of the mechanisms involved in the treatment. In general, a dosimetry model needs to reflect the complex biological processes constituting the treatment response, at the same time being simple and robust enough for clinical implementation. This section summarizes different PDT dosimetry models that have been used for preclinical and clinical PDT. The classification proposed by Wilson *et al.* [63] has been employed to categorize these models into four main classes;

1) **Explicit dosimetry:** the dose metric is based on some or all of the parameters relevant to the PDT reactions, that is, light, photosensitizer, and oxygen.
2) **Implicit dosimetry:** the dose metric is based on a single parameter, for example, the photosensitizer photobleaching kinetics, that implicitly depends on all parameters relevant for the explicit dose model.
3) **Direct dosimetry:** the dose metric is based on the total amount of singlet oxygen ($^1O_2$) produced during the treatment.
4) **Biological response:** the dose metric is related to the immediate treatment-induced biological response, such as blood flow changes.

The models listed above are more or less complex, but still only represent crude simplifications of the biological processes involved in PDT. Factors that are not taken into account are the activation of an immune response, possible synergistic effects caused by simultaneous delivery of anti-angiogenic or immune stimulating agents (Section 8.2.4), certain stress conditions affecting individual cells in the treated tissue volume, and treatment-induced re-localization of the photosensitizer. With the exception of the biological response model, they all rely on the PDT process being described by the set of reactions given in Table 8.2.

Table 8.2 Rate equations for PDT-associated reactions following light absorption; PP: photoproduct, OP: various oxidative products, A: target tissue constituents.

Process	Reaction	Rate constant	Rate equation
Light absorption	$\Phi + S_0 \rightarrow S_1$	k_f	(8.1)
Inter-system crossing	$S_1 \rightarrow T_1$	k_{ix}	(8.2)
Fluorescence	$S_1 \rightarrow S_0 + h\nu$	k_p	(8.3)
Internal conversion	$S_1 \rightarrow S_0$	k_{icS}	(8.4)
Phosphorescence	$T_1 \rightarrow S_0 + h\nu$	k_{ot}	(8.5)
Internal conversion	$T_1 \rightarrow S_0$	k_{icT}	(8.6)
Photobleaching	$S_x + {}^1O_2 \rightarrow PP + OP$	k_{osI}	(8.7)
Photobleaching	$S_x + OP \rightarrow PP + OP$	k_{osII}	(8.8)
Type I reaction	$T_1 + A \rightarrow PP + OP$	k_{ta}	(8.9)
Type II reaction	$^1O_2 + A \rightarrow OP$	k_{oa}	(8.10)
Luminescence	$^1O_2 \rightarrow {}^3O_2 + h\nu$	k_d	(8.11)
Internal conversion	$^1O_2 \rightarrow {}^3O_2$	k_{icO}	(8.12)

8.4.1
Explicit Dosimetry

The explicit PDT dosimetry model ideally takes into account the entire process from light absorption, via formation of triplet state photosensitizer to Type I and II reactions with target tissue constituents as described by rate equations (8.1)–(8.12). In its most comprehensive form it would need to consider the light distribution, $\Phi(r, t)$, the photosensitizer concentration, $[S_0(r, t)]$, as well as the oxygen level, $[^3O_2](r, t)$, on the microscopic scale. As PDT-treated tissue often displays a distinct border to unaffected tissue, explicit dose models are used in combination with the concept of a "threshold dose," that is, only tissue regions exposed to a PDT dose exceeding some threshold value, that is, $D_{PDT}(r, t) > D_{threshold}(r, t)$, experience irreversible photodamage. An explicit dosimetry model taking into account the processes described by (8.1)–(8.12) is theoretically capable of describing the complex interdependencies of all variables involved in the PDT process (Figure 8.3).

In its simplest form, clinically employed dose models normally specify only the delivered photosensitizer (in mg per kg b.w.) and light (in J cm^{-2} for topical irradiation, in J cm^{-1} for cylindrical diffusing light sources, and in J for point sources, such as cut-end or isotropic fiber tips) doses and the DLI. Thus, patient-specific photosensitizer uptake, light distribution, and target tissue geometry, for example, the thickness of a skin lesion or the location of nearby OAR, are disregarded.

Unsurprisingly, preclinical as well as clinical studies have indicated that there is a need to take into account intra- and inter-patient variations in photosensitizer and light distribution. Hence, an improved correlation between the dose metric, D_{PDT}, and the resulting PDT effect has been observed when taking into account the light dose, Eq. (8.13) [64], or the light-plus-photosensitizer dose, Eq. (8.14) [65]. The latter dose model hence takes into account the process described by rate equation (8.13). The spatial and temporal dependencies indicated in Eqs (8.13) and (8.14) emphasize the need to monitor these parameters with respect both to spatial and temporal variations:

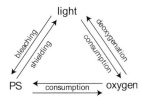

Figure 8.3 Some of the interdependencies between light, photosensitizer (PS), and oxygen. Light induces photosensitizer photobleaching and oxygen consumption. PDT-induced tissue de-oxygenation and blood volume changes affect the tissue absorbance. In addition, high photosensitizer concentration increases light absorption and thus limits light penetration. Finally, the excited photosensitizer consumes oxygen via the photodynamic reaction.

$$D_{PDT}(\boldsymbol{r}, t) \propto \int \Phi(\boldsymbol{r}, t) \mathrm{d}t \tag{8.13}$$

$$D_{PDT}(\boldsymbol{r}, t) \propto \int [S_0(\boldsymbol{r}, t)] \Phi(\boldsymbol{r}, t) \mathrm{d}t \tag{8.14}$$

The use of these simplified models is particularly tempting due to the relative ease by which the light dose can be manipulated and the existence of extensive theory for describing light propagation in biological tissue [66]. Supporting the validity of Eq. (8.14), reciprocity of light and drug doses has been observed [67–70], meaning that similar PDT effects can be achieved if varying the light and photosensitizer doses in such a way as to keep the product of the two invariant.

Although the two dose metrics proposed in Eqs (8.13) and (8.14) might be valid over limited dose ranges and for certain photosensitizers and treatment conditions, they are obviously quite crude as they ignore the effects of photosensitizer photo-bleaching [rate equations (8.7)–(8.9)] and the oxygen-dependence [rate equations (8.7) and (8.10)–(8.12)] on the resulting PDT effect. In fact, Eq. (8.14) predicts a more efficient treatment for constant photosensitizer level, an effect that is in contradiction with the implicit dosimetry model, as will be discussed in Section 8.4.2.

Foster *et al.* have devoted extensive work in setting up and solving the rate equations for the processes described in (8.1)–(8.12). This theoretical model thus takes into account PDT-induced photobleaching processes and oxygen consumption as well as spatial heterogeneity in photosensitizer and oxygen distribution. This highly explicit dose model has been employed to study different photosensitizer bleaching mechanisms in cell spheroids [71, 72] as well as *in vivo* [73], indicating that photobleaching processes are dependent on photosensitizer type, treatment conditions, and target tissue. For example, Photofrin and PpIX were reported to photobleach via the singlet oxygen-mediated path [rate equation (8.7)] whereas Nile blue selenium was degraded via non-$^1O_2$ processes [rate equations (8.8) and (8.9)] [71]. On the other hand, during *in vivo* PDT of rat skin Photofrin was reported to photobleach via both processes [73]. More recently, this model has been extended to also include the temporal and spatial oxygen distribution within and surrounding blood vessels and the blood flow kinetics [74]. Intriguingly, the singlet oxygen dose predicted by the theoretical model failed to correctly describe the resulting PDT effect *in vivo* for mTHPC-mediated therapy at different DLI [75].

Although the relevance of these theoretical models for clinical PDT might be questioned, they have provided valuable insight into the processes behind the photodynamic reactions. The observation of treatment-induced tissue hypoxia associated with the rapid oxygen consumption and vascular shut-down as discussed in Section 8.2 has been confirmed by these models [29]. Furthermore, they have been employed to develop treatment schemes that limit light-induced oxygen depletion by utilizing lower light delivery rates, thereby decreasing the instantaneous oxygen consumption of the photodynamic reaction, or by introducing irradiation fractionation where re-oxygenation of the tissue is allowed during the dark intervals (approx. seconds to minutes) [70, 76–78]. On the extreme end, mPDT, during which both photosensitizer and light are delivered at low rates for extended time periods, see also Section 8.2.4, has been investigated to increase the tumor-selectivity of the treatment [37].

8.4.2
Implicit Dosimetry

As tissue damage and photosensitizer bleaching are both induced by the same process(es) the use of the amount and rate of photosensitizer bleaching as well as the build-up of photoproducts has been investigated for monitoring PDT efficacy. Separate assessment of light, photosensitizer, and oxygen distribution, necessitated in the explicit dosimetry model discussed previously, are thus superfluous since one single variable (implicitly) yields information related to the amount of toxic substance induced by the therapeutic irradiation.

As high ROS levels induce pronounced and rapid photosensitizer bleaching, the implicit dose model actually invokes a relationship between the PDT effect and the photosensitizer concentration in direct conflict with Eq. (8.14). This equation implies a greater PDT dose for the case of no photobleaching, that is, when $[S_0(r, t)]$ remains high throughout the entire treatment, whereas the implicit approach takes into account the more pronounced photodamage of both photosensitizer and target cells that occurs during PDT within well oxygenated tissue. Both *in vitro* and *in vivo* studies provide evidence that the higher the degree and/or rate of photobleaching the better the PDT efficacy [79–83]. Alternatively, a correlation between treatment efficacy and the amount of treatment-induced photoproducts has also been observed [84].

Different models have been used to describe photosensitizer bleaching. The concept of a single photobleaching decay rate, that is, Eq. (8.15) [85, 86], has been commonly employed but is limited in that it can not incorporate the effects caused by varying fluence rates and availability of oxidative species. Basically, this limits the analysis to rate equation (8.13) in Table 8.2:

$$\frac{\delta[S_0(r,t)]}{\delta t} = -\beta\Phi[S_0(r,t)] \Rightarrow [S_0(r,t)] = [S_0(r,t_0)]\exp[-\beta\Phi(r,t)t] \tag{8.15}$$

$\sigma_{SO} = S_0$ absorption cross-section,
$\beta = \sigma_{SO}/h\nu$: commonly referred to as the photobleaching rate.

A second-order model has been employed in which the photobleaching behavior is described by Eqs (8.16) and (8.17) [87]. This model can be derived from rate equations (8.1)–(8.7) and (8.10) if assuming that all concentrations of short-lived intermediate reaction products, such as S_1, T_1, and $^1O_2$, remain approximately constant, that $[^3O_2]$ and $[A]$ are constant and unlimited, and that only $^1O_2$-mediated photosensitizer bleaching occurs:

$$\frac{\delta[S_0(r,t)]}{\delta t} = \frac{\Phi_\Delta\Phi_t\sigma_{SO}[S_0(r,t)]^2}{k_d + k_{oa}[A]} \tag{8.16}$$

$$[S_0(r,t)] = [S_0(r,t_0)]\left[1 + \frac{[S_0(r,t_0)]\Phi_\Delta\Phi_t\sigma_{SO}k_{os}}{k_d + k_{oa}[A]}\Phi(r,t)t\right]^{-1} \qquad (8.17)$$

Φ_Δ = singlet oxygen yield,
Φ_t = photosensitizer triplet state yield.

Both Eqs (8.15) and (8.17) correctly describe the feature of more rapid photobleaching for higher initial photosensitizer concentration but fail to incorporate fluence rate and oxygenation effects. Recently, spatially varying oxygen and PpIX levels have been shown to result in more complex fluorescence kinetics during ALA-mediated PDT in the rat esophagus [88]. Another reason for the detected photosensitizer fluorescence to deviate from the simple single-exponential decay might be the influence of the tissue optical properties on the detected fluorescence signal, as exemplified in Eq. (8.18) and (8.19). Although describing the photobleaching by a first-order process, as shown in Eq. (8.15), the resulting fluorescence signal can be shown to differ from a single-exponential in all cases except where $\mu_{eff}(r) = 0$, that is, where $\Phi(r,t)$ is spatially invariant [89]. This can be understood by considering the fact that deeper tissue layers are exposed to lower fluence rates and characterized by lower fluorescence escape probability, $\xi(r)$. In the case of superficial fluorescence detection, deeper tissue regions thus contribute with lower and more slowly decaying components. In this context it is also important to emphasize the risk of restricting treatment monitoring to the fluorescence kinetics only, as the measurement geometry might not be sensitive to the entire tumor volume:

$$F \propto \Phi(r,t)[S_0(r,t)] \Rightarrow F \propto \Phi(r,t)[S_0(r,t)]\exp[-\beta\Phi(r,t)t] \qquad (8.18)$$

$$F_{det} \propto \int [S_0(r,t_0)]\Phi(r,t)\exp[-\beta\Phi(r,t)t]\xi(r)dr \qquad (8.19)$$

F = photosensitizer fluorescence,
F_{det} = detected fluorescence,
$\xi(r)$ = fluorescence escape function.

8.4.3
Direct Dosimetry

Singlet-oxygen luminescence dosimetry (SOLD) has been investigated as a tool for direct PDT dosimetry. The technique relies on the assumption that the PDT effect is mainly exerted via the Type II pathway as discussed in Section 8.2. SOLD quantifies the amount of treatment-induced singlet oxygen by detecting the $^1O_2$ luminescence at 1270 nm (Figure 8.4). The relatively short lifetime of $^1O_2$ in biological media and the low probability for the radiative transition $^1\Delta_g^+ \rightarrow {}^3\Sigma_g^-(0)$ make it very challenging to detect this NIR luminescence signal and to discriminate it from background light and/or background fluorescence. However, the recently improved performance

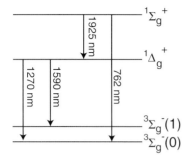

Figure 8.4 Energy level diagram of molecular oxygen illustrating different radiative decay processes together with the corresponding emission wavelengths.

of NIR-sensitive photomultiplier tubes and time-gated detection has enabled the detection of singlet-oxygen production in parallel to PDT both in *in vitro* and *in vivo* systems [90, 91]. Despite varying treatment conditions, such as different light delivery rates and photosensitizer and light doses, a strong correlation between cell survival and the cumulative $^1O_2$ luminescent count has been demonstrated both in cell suspensions and *in vivo* [90]. These observations thus support both the notion of singlet oxygen as the main cytotoxic agent and the use of the total amount of singlet oxygen as a direct and predictive PDT dose metric [92].

Notably, however, the amount of PDT-induced $^1O_2$ does not necessarily provide a universal dose metric. First, singlet oxygen dosimetry only takes into account Type II reactions and, hence, can not reflect the more macroscopic effects caused by vascular shutdown or the activation of an immune response. Second, $\int_t [^1O_2(t)]$ might be a poor dose metric if the singlet oxygen is being produced at biologically irrelevant sites as determined by the photosensitizer localization. For example, high singlet-oxygen luminescence can be detected in situations where the photosensitizer is localized within the cell cytosol. Here, the relatively low density of PDT-targets decreases the probability of luminescence quenching, effectively increasing the 1270-nm luminescence intensity. In contrast, the very short lifetime of singlet oxygen molecules being produced within membranes leads to undetectable luminescence levels. Hence, continuous-wave SOLD measurements are inherently more sensitive to the "non-useful" part of the PDT-induced $^1O_2$ luminescence. The use of the luminescence signal normalized by the singlet oxygen lifetime, or lifetime measurements, see Eq. (8.20), might, however, offer a better approach for monitoring the amount of "useful" singlet oxygen. Singlet oxygen produced at biologically relevant sites, such as the cell membrane, is characterized by a dramatically reduced lifetime compared to the 3 µs that have been observed within the cell cytosol [17, 93]. Future challenges include the use of SOLD for the micro-localization of the PDT dose in model systems as well as in the clinical, possibly interstitial setting:

$$\int L_{1270nm}(t)dt = \int \frac{^1O_2(t)}{\tau_R} dt = \frac{N\sigma_{SO}[S_0]\Phi_\Delta\tau_D}{\tau_R} \tag{8.20}$$

$\tau_R = {}^1O_2$ radiative decay lifetime,

$N = $ photons cm^{-2},

$\tau_D = {}^1O_2$ decay lifetime.

8.4.4
Biological Response

Treatment-induced physiological alterations of the tissue also offer the possibility to monitor the PDT effect in real-time. For example, the vascular shutdown, important for VTP (Section 8.2), has been monitored by means of diffuse correlation spectroscopy (DCS) [94], laser Doppler perfusion measurements [95], interstitial Doppler optical coherence tomography (DOCT) [96], Doppler ultrasound [97], fluorescein angiography [98], and blood oxygenation level dependent (BOLD) MRI [99]. Furthermore, electrical impedance spectroscopy (EIS) has been employed to study the tissue permittivity, related to tissue edema, cell necrosis, and vascular damage, during PDT [100]. Other methods that can provide information related to cell damage are high-frequency ultrasound [101], PET [102], diffusion-weighted magnetic resonance imaging (DW-MRI) [103], and bioluminescence imaging [104].

8.4.5
Summary of PDT Dose Models

Table 8.3 summarizes some advantages and drawbacks of the dose models presented.

Table 8.3 Advantages and disadvantages of the PDT dosimetry models.

Model	Advantages	Disadvantages
Explicit	Conceptually straightforward, simplified dosimetry (Φ and/or PS) can be used in some settings	Need to monitor multiple parameters, D_{PDT} as a function of Φ. S_0 and $^3O_2$ is unknown, absolute measurements might be required, does not consider vascular effects, does not consider immunological effects
Implicit	Technically easy and cost-effective, dimensionless parameters, for example, $F(t)/F(t_0)$ can be used	Mechanism unknown, dependent on optical properties, does not consider vascular effects, does not consider immunological effects, technically more complex if also including monitoring of tissue optical properties
Direct	Direct tool via monitoring of the active component	Technically challenging, expensive equipment, sensitive to "inactive" $^1O_2$ molecules, does not consider vascular effects, does not consider immunological effects
Biological response	Direct tool, in particular for vascular-targeted PDT (VPT), can relatively easily be extended to yield information on tissue oxygenation	Requires validation, indication specific, does not consider immunological effects

8.5
Clinical Implementation

Since first coining the term "photodynamic therapy" more than 100 years ago [1, 2], PDT has acquired clinical approvals for the treatment or palliation of some malignant and non-malignant indications, see Table 8.1, and many more trials are, or have been, investigating its use for numerous other indications, see, for example, http://www.clinicaltrials.gov. In contrast to radiation therapy, PDT can be repeated without cumulative toxicity. PDT might thus play an important role for recurrent tumors as these patients are often not eligible for further radiation therapy or surgery. Additional advantages of PDT include preservation of the tissue structure, good cosmetic outcome, and cost-effectiveness. In contrast to radiotherapy and the use of some cytotoxic drugs, PDT induces only a limited effect on DNA and therefore is not considered cancerogenic. The use of targeted light delivery and photosensitizer with selective uptake in the target tissue promotes treatment selectivity.

Despite the successful use of PDT for some indications, PDT dosimetry is in large under-developed. Table 8.4 lists currently active trials focusing on PDT dosimetry. As can be seen, clinical PDT is still mostly based upon empirical light and photosensitizer dose escalation trials without much consideration of the intra- and

Table 8.4 Registered PDT dosimetry trials (http://www.clinicaltrials.gov, June 2009). Search terms: photodynamic therapy & dose & recruiting. D_L: light dose, esc.: dose escalation.

Indication	Photosensitizer	Aim
AMD	Verteporfin	Phase IV: varying D_L, combination with ranibizumab
	Verteporfin	Phase II: varying D_L, combination with ranibizumab
	Verteporfin	Phase II: combination with Fluocinolone acetonide/Medidur
NMSM	ALA	Fluor. imaging of PpIX distr.
Warts	ALA	Phase III: fluor. imaging of PpIX distr.
Psoriasis	PUVA (Psoralen plus UV-A) + Acitretin or fumaric acid ester	Phase III: combination synergy
Chronic leg ulcer	PPA904	Phase II: repeat treatment
Oral leukoplakia	ALA	Phase I/II: D_L & [PS] esc.
Oral cavity	HPPH	Phase I: D_L esc., fluor. imaging
Lung	HPPH	Phase I: [PS] esc.
Larynx	HPPH	Phase I: D_L esc.
H&N	ALA	Phase I: D_L esc.
	HPPH	Phase I: D_L esc.
Breast	Photofrin	Phase I: D_L esc.
Brain	Verteporfin	Phase I: [PS] esc.
Neurofibroma	HPPH	Phase I: D_L & [PS] esc.

inter-patient variations and without the aim of monitoring the immediate response to PDT. This section gives a brief overview of clinical PDT dosimetry, ranging from approved protocols to feasibility studies in preclinical settings.

PDT for the treatment of AMD and other eye diseases related to choroidal neovascularization (CNV) relies on the VTP approach and contains recommendations for the total delivered light and drug doses as well as for the irradiance (W cm^{-2}). As discussed in Section 8.2.4, the combination of PDT, relying on Visudyne® and pharmaceutics containing VEGF inhibitors and anti-inflammatory steroids [105], has been shown to improve treatment outcomes. The observed synergy is attractive as lowered light and photosensitizer doses can be employed, in some cases resulting in improved treatment selectivity [106]. The use of two-photon PDT has so far only been investigated preclinically but might provide a potential for highly selective treatment of CNV [107].

Partly due to the relative ease of light delivery, PDT for superficially located malignancies, such as skin cancer and tumors within hollow organs, has come far in terms of clinically approved protocols (see Table 8.1). For example, in 1993 the first clinical PDT approval was issued where Photofrin was approved for the treatment of superficial papillary bladder cancer. PDT has shown promising treatment results for esophageal cancers and Barrett's high-grade dysplasia [108, 109] and constitutes an interesting alternative to surgery and radiation therapy for head and neck cancers due to its greatly reduced morbidity and disfigurement [110, 111]. Although not yet clinically approved, Photofrin-PDT for cholangiocarcinoma has shown very promising treatment results [112] and an on-going Phase III trial, PHOTOSTENT 2, aims at further investigating porfimer sodium-mediated PDT for advanced or metastatic cholangiocarcinomas and other biliary tract tumors. Based on observed treatment outcomes and occurrence of side-effects, the conventionally employed dosimetry models are limited to only specifying administered light and drug doses. For example, 20 J cm^{-2} and 0.15 mg kg^{-1} are used for mTHPC-mediated PDT [113] and 50–100 J cm^{-2} and 2 mg kg^{-1} for Photofrin-mediated [110] PDT of head and neck cancer. PDT of NMSM employs 37 J cm^{-2} at 635 nm when utilizing Metvix® and 10 J cm^{-2} at 400–450 nm for Levulan Kerastick®.

Light applicators for superficial irradiation include bare-ended fibers, microlens-equipped fibers, and diffusing fibers with isotropic, cylindrical or balloon-equipped tips. Such light applicators can easily be inserted via the working channel of most endoscopes. PDT dosimetry is, however, highly under-developed in most clinical settings, as was also highlighted in the AAPM Task Group Report on PDT from 2005 [114].[1] Fundamental measures such as shielding surrounding tissue from light or employing green excitation light, thus limiting light penetration for the treatment of superficial malignancies, are sometimes used clinically. The actual light dose, as determined by the tissue- and patient-specific optical properties, is not known. The deposited light dose decreases dramatically with depth and is also influenced by

1) General Medical Physics Committee of the Science Council Task Group 5, Report No. 88, published for the American Association of Physicists in Medicine by Medical Physics Publishing (AAPM), www. medicalphysics.org.

the so-called fluence rate build-up factor. This effect, caused by light reflection at the tissue–air interface, can effectively increase the surface irradiance within hollow organs as well as the fluence rate just below the surface [64]. The concept of individualized, real-time light dosimetry is currently being investigated for naso-pharyngeal cancer [115], including monitoring of fluence rate distribution and build-up factor, and for intraperitoneal cancer [116], incorporating measurements of tissue optical properties, photosensitizer uptake, and tissue oxygenation.

Attempts at modeling the photosensitizer depth distribution following topical application have been made with the aim to implement a "light-plus-photosensitizer" dosimetry, see Eq. (8.14) in Section 8.4.1. For ALA-induced PpIX these models need to take into account the diffusion of ALA molecules, their uptake into cells as well as their conversion into PpIX [117, 118]. Although of limited clinical usefulness, such models might be relevant when trying to understand differences in tissue penetration of ALA and its esters, the importance of increasing skin permeability, for example, by removing the stratum corneum, and the PpIX kinetics, including its reappearance during dark cycles ($\sim$ hours) [119].

In the case of highly tumor-selective photosensitizer accumulation, as obtained, for example, in NMSM, malignant gliomas [120], and bladder malignancies [121] following ALA or h-ALA application, an accurate light dosimetry is of secondary importance.[2] In such cases simplified dose models, for example, aiming at achieving complete photobleaching or relying on the implicit dosimetry model as discussed in Section 8.4.2, can be employed. Fluorescence imaging in parallel to PDT for NMSM has revealed an inverse relationship between the photobleaching degree and the fluence rate at the therapeutic wavelength [80, 82] as well as a correlation between the photobleaching rate and the initial treatment outcome [82]. Real-time spectroscopic monitoring has the potential to improve treatment outcome by identifying situations where too high fluence rates induce oxygen depletion or where complete photosen-sitizer consumption has been achieved. The risk with such spectroscopic monitor-ing, however, is its low sensitivity for deeper tissue regions and hence care should be taken to administer a light dose sufficient to also target deeper, hopefully photo-sensitized, tissue regions.

SOLD measurements have recently been brought into human *in vivo* use where a fiber-based system has been employed for the time-resolved detection of $^1O_2$ luminescence during PDT of normal skin [122]. In agreement with the direct PDT dosimetry model, a correlation could be observed between the singlet oxygen production and the treatment-induced erythema and edema. Fiber-based singlet oxygen detection setups might hold potential for PDT dosimetry within hollow or solid organs but the low luminescence intensity still constitutes a challenging problem for interstitial applications.

Interstitial PDT has been investigated for targeting deeply lying tumors, for example, in the prostate [123–125], liver [126], and brain [120], and massive tumors

---

2) In 2007, Gliolan, was approved for fluorescence-guided resection (FGR) of residual malignant glioma. In 2005, Hexvix obtained approval for fluorescence-mediated diagnosis of bladder carcinoma.

of the skin [127] or in the head and neck region [110, 128]. In particular, prostate-PDT has been combined with real-time spectroscopic measurements aiming at investigating various dosimetric approaches. Phase I/II studies on prostate-PDT have employed motexafin lutetium, mTHPC, WST09, or WST11. The bacteriopheophorbides, WST09 and WST11, are combined with the VTP approach. Light applicators include cylindrical diffusors or bare-ended fibers. Knowledge of the light and photosensitizer distribution is obtained by monitoring the light transmission between the fibers used for the therapeutic light delivery [123] or by also employing additional, isotropically emitting/detecting optical fibers that are either stationary [124] or mechanically translated within the target tissue [125, 129]. Significant intra- and inter-patient variability in tissue optical properties and photosensitizer uptake has been observed, emphasizing the need for individualized treatment parameters. Three-dimensional virtual maps of the fluence rate [130] and photosensitizer [131] distribution have also been created. Such information might thus be used for adjusting irradiation times and output powers of individual light sources to deliver a sufficient PDT dose according to either an explicit or implicit dose model.

The trend towards real-time treatment monitoring holds great promise in optimizing treatment parameters as well as understanding the immediate response of the tissue to the treatment. Aiming at a dose model based on the biological response to prostate-PDT, tissue oxygenation levels have been assessed via absorption spectroscopy in the NIR wavelength range [94, 124, 125] and the blood flow, of particular importance to VTP, has been monitored via diffuse correlation spectroscopy (DCS) [94]. For skin-PDT, blood flow changes have been monitored via Doppler shifted reflectance [132], NIR reflectance [133], and speckle imaging [134].

8.6
Where is PDT Heading?

Important issues to address in trying to assess the future of PDT are the reasons for the still relatively limited clinical use of PDT (considering its potential), whether these reasons are valid also for clinical applications yet unexplored, and if one can overcome these obstacles. Is the main reason for the few approved indications unsatisfactory treatment results, intolerable side effects, or only due to challenges in the approval process as both "drug" and "device" are needed? Certainly, all of the above-mentioned reasons are important and application-dependent. There still exists a need for a better photosensitizer capable of providing effective treatment results with tolerable side effects. Unfortunately, commercial interests are delaying the development, as companies are frequently unwilling to jeopardize a business case by using their drug for more than one indication. To improve treatment outcome and minimize side-effects one also needs to better understand the treatment mechanisms and improve the treatment dosimetry. The often crude dosimetry models currently in clinical use need to be advanced, starting by elucidating the relevance of existing dose models for different indications and identifying those situations where PDT dosimetry actually could present an improvement.

To give a clear answer to the question of where PDT is heading is indeed a difficult task. Here, we provide some indications for the future of PDT by outlining some recent trends in the PDT field. In doing so, we categorize these trends in terms of novel applications, novel light delivery modes, novel photosensitizer, and novel dosimetry, while elaborating on those aspects that can influence the treatment of solid tumors. It is also clear from the above presentation that no one single implementation of PDT will suit all types of clinical applications, but rather one has to look into optimizing the treatment for each indication. For example, PDT of prostate cancer or Barrett's esophagus presents two completely different situations; in the first case one would like to target a large tissue volume and hence need to maximize light penetration, whereas the very superficial lesions present for Barrett's esophagus requires limited treatment depths to avoid undesired damage of underlying healthy tissue.

8.6.1
Novel Applications

As a treatment modality is becoming increasingly understood and accepted and treatment parameters, photosensitizer, and dosimetry are developing, new indications may emerge. Since its first clinical use in the twentieth century, PDT has been applied to almost every bodily organ. Despite a lack of more detailed dosimetry, several indications have obtained clinical approval, see Table 8.1, and many more are currently being identified and explored, see for instance References [15, 135–138]. PDT dosimetry is being developed, as discussed in Section 8.5 (in particular Table 8.4). The success of PDT for the treatment of solid tumors might very well critically depend on the use of adequate dosimetry as provided by real-time treatment monitoring. Spectroscopic measurements performed before and during the treatment hold the potential to optimize the outcome and minimize effects of OAR (Section 8.6.4).

Here, we also point out the use of PDT as one part of a combination therapy. Many modalities for the treatment of malignant disease rely on a multi-modality approach, with the intention to improve treatment outcome and limit side-effects. For PDT, one has identified combinations, for example, the parallel administration of anti-angiogenetic- and/or anti-inflammatory drugs, that provide synergetic effects, see also Sections 8.2.4 and 8.5 and Table 8.4. Although many problems and challenges remain, promising results in preclinical and clinical settings have been seen in PDT and immune manipulation. The consequences of immunoadjuvant therapy on the PDT light and photosensitizer dosimetry is, however, yet unknown. Future studies will assess the possibility to employ lower total light and drug dosages, as well as the potential for improved selectivity when also activating the adaptive immune system [39].

There are several applications where PDT is not directly employed to eradicate malignancies. Here, we would like to point out two such indications. Firstly, extensive research has investigated the use of PDT for the treatment of microbial infections, where alternative treatment options are sought due to the increasing frequency of methicillin-resistant *Staphylococci* strains (MRSA) infections. Brown *et al.* are presently conducting a phase II multi-center randomized and placebo-controlled study on antimicrobial PDT for chronic leg ulcers and chronic diabetic foot ulcers with very promising results. An interesting review of the field was published recently [139].

Secondly, a very interesting approach is the treatment of malignant tumors with photochemical internalization (PCI). This concept, developed by Berg *et al.*, employs sensitized vesicles that can enter cells via endocytosis. These vesicles are exposed to activating light, whereby the photodynamic action leads to damage of the photo-sensitized membrane and their content can be released (or internalized) into the cytosol of the cell [140, 141]. The vesicles could be loaded with a toxic agent, for instance bleomycin, thus selectively destroying irradiated cells. This allows a highly selective chemotherapeutic treatment, potentially resulting in much less side-effects. This could be very interesting for several different clinical applications, and might also be fruitfully combined with the optical feedback techniques described in Reference [123]. Similarly, Madsen *et al.* have employed the same concept to temporarily open the blood–brain barrier (BBB) to deliver therapeutic drugs into the brain tumor [142]. This can be a very powerful technique to minimize recurrences following therapy by targeting also those malignant cells that have already migrated into the healthy brain and that are normally protected by an intact BBB. Clearly, light dosimetry is of upmost importance for this application.

8.6.2
Novel Light Delivery Modes

We mention here two different types of novel and interesting light delivery modes: two-photon and metronomic PDT (mPDT).

Two-photon PDT has been investigated due to the possibility of using long excitation wavelengths, thus potentially providing increased tissue penetration. The principle of two-photon PDT is very similar to that of two-photon microscopy, where out of focus areas are not excited due to the quadratic dependence on the photon density. It thus allows depth selective very local treatment – see for instance Reference [107]. As the two-photon cross section is very small, it has been convincingly argued, however, that this excitation mode does not provide any advantage for PDT of thick lesions. Instead, two-photon PDT might be interesting for applications requiring highly localized therapy, as for AMD and infiltrating micro-metastases.

For mPDT, see also Section 8.2.4, light and photosensitizer are delivered over extended time periods. Owing to the extended time consumption, this treatment modality might be more attractive for patients already requiring long and costly hospitalization periods. Technically, small portable light and drug sources need to be implemented and the requirements of accurate light dosimetry need to be elucidated.

8.6.3
Novel Photosensitizer Development

There is still a strong interest in and need for new and better PDT-photosensitizers. Besides the aim to obtain increased treatment efficacy and minimize side effects, one also wishes to circumvent the commercial block presently hindering the use of existing drugs for new indications. In the drug development process it has also become evident that it is difficult to identify one chemical molecule possessing all the optimal properties for a PDT treatment. Rather, one has to exploit more complex structures that can provide the possibility of adjusting certain functions independent of the photosensitizer agent. The various characteristics can thus be associated with different parts of the of the photosensitizer complex. Increasing interest can be seen in more complex agents that explore PDT mechanisms at the molecular level and that also enable the differentiation of the targeting ability and the photophysical properties of importance once the drug reached the target. A method often proposed is the encapsulation of the photosensitizer into a liposome, thereby separating the photophysical properties (determined by the photosensitizer molecule) and the pharmacokinetic properties (determined by the liposome) [143, 144]. This concept permits the use a very potent photosensitizer, for example, mTHPC, while modifying the poor pharmacokinetic properties of the drug [57, 145]. Another alternative is linking the photosensitizer to an antibody [146, 147], that is, working within the concept of photoimmunotherapy (PIT) as discussed in Section 8.2.4. This alternative may be very attractive, as antibodies are readily available for many tumors. The challenges are to obtain a concentration sufficient to eradicate all tumor cells and to obtain a high selectivity with respect to normal surrounding cells. Another interesting and recent development towards increased drug selectivity is the use of activatable photosensitizers, also referred to as photosensitizer beacons. These have a configuration resulting in inactivation of the photosensitizing ability by a molecular quenching mechanism. This molecular configuration alters once the complex binds to a specific target molecule, where it changes configuration and the quenching is destroyed. This means that the photosensitizer is only efficient once it is bound to a specific target in the tissue [148–150]. These approaches offer an improved selectivity as compared to most currently employed photosensitizers and hence would pose less severe demands in terms of accurate light dosimetry. See Section 8.2.4 for further discussion on photosensitizer targeting.

Nanoparticle-based photosensitizers have also become interesting in striving to develop photosensitizers that can be more easily engineered to optimize all the different characteristics required for successful PDT. One approach is based on the incorporation into or binding of the photosensitizer to a nanoparticle, where the nanoparticle is used as targeting agent. This can be achieved by employing, for instance, gold or silica nanoparticles decorated with antibodies or peptide sequences [151]. A slightly different approach involves enclosing the photosensitizer into a liposome and coating the liposome with particles providing the targeting properties. Yet another alternative is to employ photoactive nanoparticles, in which case the photophysical properties of the nanoparticles are also exploited. For instance, porous silicon particles can absorb light and react with oxygen to generate singlet oxygen [152]. The ease by which the absorption band of the nanoparticle can be tailored opens up the possibilities to employ either a FRET process for exciting the attached photosensitizer [153] or long excitation wavelengths, characterized by deeper penetration into tissue, that are absorbed by the nanoparticle followed by energy transfer to the coupled photosensitizer [154, 155]. It would in this way be possible to treat deeper lying and localized lesions from outside the body. Most of the approaches based on nanoparticles are at a very early phase, and toxicity issues still need to be investigated.

8.6.4
Novel Implementation of Dosimetry and Dosimetric Measurements

Dosimetry aspects have been the main interest of this chapter. From the dosimetric perspective, significant development has been accomplished during the last couple of years, providing a much better base for further development of PDT for new indications, especially for treatment of solid tissue volumes. Hence, improved understanding of the mechanisms behind the PDT response and extended dosimetry models reflecting these processes have emerged. Although clinical routine still relies on relatively crude dose models, for example, the simple threshold dose model [Eq. (8.13) in Section 8.4.1], the trend is towards including more parameters, for example, the "light-and-photosensitizer" threshold dose model [Eq. (8.13) in Section 8.4.1]. This necessarily requires more input parameters, in particular when also attempting individualized treatment planning and PDT dosimetry. There is a pronounced development towards real-time/online PDT-treatment monitoring (Section 8.4). Spectroscopic techniques are often employed for these measurements, where the use of small-diameter fiber-based probes limits invasiveness. This aspect has been further advanced by the use of the same fibers for both therapeutic irradiation and spectroscopic measurements [127, 142, 156–158]. This concept was developed for interstitial PDT relying on an explicit dosimetry model, where multiple bare-ended fibers are employed. Initial clinical results on light and photosensitizer three-dimensional distribution have been published recently [131]. This technique may become important for the development of PDT for bigger tissue volumes requiring many fibers for an efficient light delivery.

The notion of real-time treatment monitoring is attractive also when applying an implicit dosimetry model or a dose model based on the biological response to the PDT-treatment. Here we take the opportunity to also mention some exciting new monitoring techniques that might prove useful for PDT monitoring: optical coherence tomography (OCT) might be employed to monitor scattering changes during PDT or combined with fluorescence measurements to allow more quantitative photosensitizer monitoring, Doppler-OCT or opto-acoustic tomography might be employed before, during, and after PDT to assess vascular shut-down. The dose model based on the biological response to PDT also offers the possibility to gain additional insight into the mechanism behind the resulting tissue damage. Such information might prove essential when attempting synergistic effects for combination therapies (Section 8.2.4). As a summary, we again emphasize the possibility offered by real-time dosimetry and/or combination strategies to diminish treatment-related side effects as treatment parameters can be individualized and/or each treatment modality must not be used to any extreme doses.

References

1 von Tappeiner, H. and Jodlbauer, A. (1907) Die sensibilisierende Wirkung fluorescierender Substanzen, in *Gesammelt Untersuchungen über die photodynamische Erscheinung*, FCW Vogel, Leipzig.

2 von Tappeiner, H. and Jodlbauer, A. (1904) Über wirkung der photodynamischen (fluorieszierenden) stoffe auf protozoan und enzyme. *Dtsch. Arch. Klin. Med.*, **80**, 427–487.

3 Dougherty, T.J., Kaufman, J.E., Goldfarb, A., Weishaupt, K.R., Boyle, D., and Mittleman, A. (1978) Photoradiation therapy for the treatment of malignant tumors. *Cancer Res.*, **38** (8), 2628–2635.

4 Chakravarthy, U., Soubrane, G., Bandello, F., Chong, V., Creuzot-Garcher, C., Dimitrakos, S.A., Korobelnik, J.F., Larsen, M., Mones, J., Pauleikhoff, D., Pournaras, C.J., Staurenghi, G., Virgili, G., and Wolf, S. (2006) Evolving European guidance on the medical management of neovascular age related macular degeneration. *Br. J. Ophthalmol.*, **90** (9), 1188–1196.

5 Garcia-Zuazaga, J., Cooper, K.D., and Baron, E.D. (2005) Photodynamic therapy in dermatology: current concepts in the treatment of skin cancer. *Expert Rev. Anticancer. Ther.*, **5** (5), 791–800.

6 Klein, A., Babilas, P., Karrer, S., Landthaler, M., and Szeimies, R.M. (2008) Photodynamic therapy in dermatology – an update 2008. *J. Dtsch. Dermatol. Ges.*, **6** (10), 839–845.

7 Muller, P.J. and Wilson, B.C. (1990) Photodynamic therapy of malignant brain tumours. *Can. J. Neurol. Sci.*, **17** (2), 193–198.

8 Whitehurst, C., Pantelides, M.L., Moore, J.V., Brooman, P.J., and Blacklock, N.J. (1994) *In vivo* laser light distribution in human prostatic carcinoma. *J. Urol.*, **151** (5), 1411–1415.

9 Moore, C.M., Pendse, D., and Emberton, M. (2009) Photodynamic therapy for prostate cancer – a review of current status and future promise. *Nat. Clin. Pract. Urol.*, **6** (1), 18–30.

10 Huang, Z., Xu, H.P., Meyers, A.D., Musani, A.I., Wang, L.W., Tagg, R., Barqawi, A.B., and Chen, Y.K. (2008) Photodynamic therapy for treatment of solid tumors – potential and technical challenges. *Technol. Cancer Res. Treatment*, **7** (4), 309–320.

11 Wilson, B.C. and Patterson, M.S. (2008) The physics, biophysics and technology of photodynamic therapy. *Phys. Med. Biol.*, **53** (9), R61–R109.

12 Zhu, T.C. and Finlay, J.C. (2008) The role of photodynamic therapy (pdt) physics. *Med. Phys.*, **35** (7), 3127–3136.

13 Ochsner, M. (1997) Photophysical and photobiological processes in the photodynamic therapy of tumours. *J. Photochem. Photobiol. B*, **39** (1), 1–18.

14 Weishaupt, K.R., Gomer, C.J., and Dougherty, T.J. (1976) Identification of singlet oxygen as the cytotoxic agent in photoinactivation of a murine tumor. *Cancer Res.*, **36** (7 Pt 1), 2326–2329.

15 Dolmans, D.E.J.G., Fukumura, D., and Jain, R.K. (2003) Photodynamic therapy for cancer. *Nat. Rev. Cancer*, **3** (5), 380–387.

16 Henderson, B.W. and Fingar, V.H. (1987) Relationship of tumor hypoxia and response to photodynamic treatment in an experimental mouse tumor. *Cancer Res.*, **47** (12), 3110–3114.

17 Kuimova, M.K., Yahioglu, G., and Ogilby, P.R. (2009) Singlet oxygen in a cell: spatially dependent lifetimes and quenching rate constants. *J. Am. Chem. Soc.*, **131** (1), 332–340.

18 Oleinick, N.L., Morris, R.L., and Belichenko, I. (2002) The role of apoptosis in response to photodynamic therapy: what, where, why, and how. *Photochem. Photobiol. Sci.* **1** (1), 1–21.

19 Lilge, L., Portnoy, M., and Wilson, B.C. (2000) Apoptosis induced in vivo by photodynamic therapy in normal brain and intracranial tumour tissue. *Br. J. Cancer*, **83** (8), 1110–1117.

20 Almeida, R.D., Manadas, B.J., Carvalho, A.P., and Duarte, C.B. (2004) Intracellular signaling mechanisms in

photodynamic therapy. *Biochim. Biophys. Acta*, **1704** (2), 59–86.

21 van Duijnhoven, F.H., Aalbers, R.I., Rovers, J.P., Terpstra, O.T., and Kuppen, P.J. (2003) The immunological consequences of photodynamic treatment of cancer, a literature review. *Immunobiology*, **207** (2), 105–113.

22 Moor, A.C. (2000) Signaling pathways in cell death and survival after photodynamic therapy. *J. Photochem. Photobiol. B*, **57** (1), 1–13.

23 Dahle, J., Bagdonas, S., Kaalhus, O., Olsen, G., Steen, H.B., and Moan, J. (2000) The bystander effect in photodynamic inactivation of cells. *Biochim. Biophys. Acta*, **1475** (3), 273–280.

24 Kessel, D. and Oleinick, N.L. (2009) Initiation of autophagy by photodynamic therapy. *Methods Enzymol.*, **453**, 1–16.

25 Henderson, B.W. and Dougherty, T.J. (1992) How does photodynamic therapy work? *Photochem. Photobiol.*, **55** (1), 145–157.

26 Fingar, V.H. (1996) Vascular effects of photodynamic therapy. *J. Clin. Laser Med. Surg.*, **14** (5), 323–328.

27 Chen, B., Pogue, B.W., Hoopes, P.J., and Hasan, T. (2006) Vascular and cellular targeting for photodynamic therapy. *Crit. Rev. Eukaryot. Gene Expr.*, **16** (4), 279–305.

28 Fingar, V.H., Kik, P.K., Haydon, P.S., Cerrito, P.B., Tseng, M., Abang, E., and Wieman, T.J. (1999) Analysis of acute vascular damage after photodynamic therapy using benzoporphyrin derivative (bpd). *Br. J. Cancer*, **79** (11–12), 1702–1708.

29 Busch, T.M., Wileyto, E.P., Emanuele, M.J., Del Piero, F., Marconato, L., Glatstein, E., and Koch, C.J. (2002) Photodynamic therapy creates fluence rate-dependent gradients in the intratumoral spatial distribution of oxygen. *Cancer Res*, **62** (24), 7273–7279.

30 Castano, A.P., Mroz, P., and Hamblin, M.R. (2006) Photodynamic therapy and anti-tumour immunity. *Nat. Rev. Cancer*, **6** (7), 535–545.

31 Korbelik, M. (2006) Pdt-associated host response and its role in the therapy outcome. *Lasers Surg. Med.*, **38** (5), 500–508.

32 Korbelik, M. and Dougherty, T.J. (1999) Photodynamic therapy-mediated immune response against subcutaneous mouse tumors. *Cancer Res.*, **59**, 1941–1946.

33 Korbelik, M., Krosl, G., Krosl, J., and Dougherty, G.J. (1996) The role of host lymphoid populations in the response of mouse emt6 tumor to photodynamic therapy. *Cancer Res.*, **56** (24), 5647–5652.

34 Gollnick, S.O., Vaughan, L., and Henderson, B.W. (2002) Generation of effective antitumor vaccines using photodynamic therapy. *Cancer Res.*, **62** (6), 1604–1608.

35 Dougherty, T.J., Gomer, C.J., Henderson, B.W., Jori, G., Kessel, D., Korbelik, M., Moan, J., and Peng, Q. (1998) Photodynamic therapy. *J. Natl. Cancer Inst.*, **90** (12), 889–905.

36 Henderson, B.W., Busch, T.M., and Snyder, J.W. (2006) Fluence rate as a modulator of pdt mechanisms. *Lasers Surg. Med.*, **38** (5), 489–493.

37 Bisland, S.K., Lilge, L., Lin, A., Rusnov, R., and Wilson, B.C. (2004) Metronomic photodynamic therapy as a new paradigm for photodynamic therapy: rationale and preclinical evaluation of technical feasibility for treating malignant brain tumors. *Photochem. Photobiol.*, **80** (22–30), 22–30.

38 Dolmans, D.E., Kadambi, A., Hill, J.S., Flores, K.R., Gerber, J.N., Walker, J.P., Rinkes, I.Borel, Jain, R.K., and Fukumura, D. (2002) Targeting tumor vasculature and cancer cells in orthotopic breast tumor by fractionated photosensitizer dosing photodynamic therapy. *Cancer Res.*, **62** (15), 4289–4294.

39 Zuluaga, M.F. and Lange, N. (2008) Combination of photodynamic therapy with anti-cancer agents. *Curr. Med. Chem.*, **15** (17), 1655–1673.

40 Gomer, C.J., Ferrario, A., Luna, M., Rucker, N., and Wong, S. (2006) Photodynamic therapy: combined modality approaches targeting the tumor microenvironment. *Lasers Surg. Med.*, **38** (5), 516–521.

41 Kosharskyy, B., Solban, N., Chang, S.K., Rizvi, I., Chang, Y., and Hasan, T. (2006) A mechanism-based combination

therapy reduces local tumor growth and metastasis in an orthotopic model of prostate cancer. *Cancer Res.*, **66** (22), 10953–10958.

42 Agostinis, P., Buytaert, E., Breyssens, H., and Hendrickx, N. (2004) Regulatory pathways in photodynamic therapy induced apoptosis. *Photochem. Photobiol. Sci.*, **3** (8), 721–729.

43 Dadgostar, H. and Waheed, N. (2008) The evolving role of vascular endothelial growth factor inhibitors in the treatment of neovascular age-related macular degeneration. *Eye*, **22** (6), 761–767.

44 Qiang, Y.G., Yow, C.M., and Huang, Z. (2008) Combination of photodynamic therapy and immunomodulation: current status and future trends. *Med. Res. Rev.*, **28** (4), 632–644.

45 Kwitniewski, M., Juzeniene, A., Glosnicka, R., and Moan, J. (2008) Immunotherapy: a way to improve the therapeutic outcome of photodynamic therapy? *Photochem. Photobiol. Sci.*, **7** (9), 1011–1017.

46 Winters, U., Daayana, S., Lear, J.T., Tomlinson, A.E., Elkord, E., Stern, P.L., and Kitchener, H.C. (2008) Clinical and immunologic results of a phase ii trial of sequential imiquimod and photodynamic therapy for vulval intraepithelial neoplasia. *Clin. Cancer Res.*, **14** (16), 5292–5299.

47 Wang, Xiu Li, Wang, Hong Wei, Guo, Ming Xia, and Huang, Zheng (2007) Combination of immunotherapy and photodynamic therapy in the treatment of bowenoid papulosis. *Photodiag. Photodyn. Ther.*, **4** (2), 88–93.

48 Szygula, M., Pietrusa, A., Adamek, M., Wojciechowski, B., Kawczyk-Krupka, A., Cebula, W., Duda, W., and Sieron, A. (2004) Combined treatment of urinary bladder cancer with the use of photodynamic therapy (pdt) and subsequent bcg-therapy: a pilot study. *Photodiag. Photodyn. Ther.*, **1** (3), 241–246.

49 Garbo, G.M., Vicente, M.G., Fingar, V., and Kessel, D. (2003) Effects of ursodeoxycholic acid on photodynamic therapy in a murine tumor model. *Photochem. Photobiol.*, **78** (4), 407–410.

50 Miller, J.D., Baron, E.D., Scull, H., Hsia, A., Berlin, J.C., McCormick, T., Colussi, V., Kenney, M.E., Cooper, K.D., and Oleinick, N.L. (2007) Photodynamic therapy with the phthalocyanine photosensitizer pc 4: the case experience with preclinical mechanistic and early clinical-translational studies. *Toxicol. Appl. Pharmacol.*, **224** (3), 290–299.

51 Nyman, E.S. and Hynninen, P.H. (2004) Research advances in the use of tetrapyrrolic photosensitizers for photodynamic therapy. *J. Photochem. Photobiol. B*, **73** (1–2), 1–28.

52 Henderson, B.W. and Bellnier, D.A. (1989) Tissue localization of photosensitizers and the mechanism of photodynamic tissue destruction. *Ciba Found. Symp.*, **146**, 112–125, discussion 125–130.

53 Peng, Q., Soler, A.M., Warloe, T., Nesland, J.M., and Giercksky, K.E. (2001) Selective distribution of porphyrins in skin thick basal cell carcinoma after topical application of methyl 5-aminolevulinate. *J. Photochem. Photobiol. B*, **62** (3), 140–145.

54 Casas, A. and Batlle, A. (2006) Aminolevulinic acid derivatives and liposome delivery as strategies for improving 5-aminolevulinic acid-mediated photodynamic therapy. *Curr. Med. Chem.*, **13** (10), 1157–1168.

55 Braathen, L.R., Szeimies, R.M., Basset-Seguin, N., Bissonnette, R., Foley, P., Pariser, D., Roelandts, R., Wennberg, A.M., and Morton, C.A. (2007) Guidelines on the use of photodynamic therapy for nonmelanoma skin cancer: an international consensus. International society for photodynamic therapy in dermatology, 2005. *J. Am. Acad. Dermatol.*, **56** (1), 125–143.

56 Morton, C.A., McKenna, K.E., and Rhodes, L.E. (2008) Guidelines for topical photodynamic therapy: update. *Br. J. Dermatol.*, **159** (6), 1245–1266.

57 Johansson, A., Svensson, J., Bendsoe, N., Svanberg, K., Alexandratou, E., Kyriazi, M., Yova, D., Gräfe, S., Trebst, T., and Andersson-Engels, S. (2007) Fluorescence and absorption assessment of a lipid mthpc formulation following

topical application in a non-melanotic skin tumor model. *J. Biomed. Opt.*, **12** (3), 034026.

58 Peng, Q., Moan, J., Nesland, J.M., and Rimington, C. (1990) Aluminum phthalocyanines with asymmetrical lower sulfonation and with symmetrical higher sulfonation: a comparison of localizing and photosensitizing mechanism in human tumor lox xenografts. *Int. J. Cancer*, **46** (4), 719–726.

59 Boyle, R.W. and Dolphin, D. (1996) Structure and biodistribution relationships of photodynamic sensitizers. *Photochem. Photobiol.*, **64** (3), 469–485.

60 Jori, G. and Reddi, E. (1993) The role of lipoproteins in the delivery of tumour-targeting photosensitizers. *Int. J. Biochem.*, **25** (10), 1369–1375.

61 Derycke, A.S. and De Witte, P.A. (2004) Liposomes for photodynamic therapy. *Adv. Drug Deliv. Rev.*, **56** (1), 17–30.

62 Konan, Y.N., Gurny, R., and Allemann, E. (2002) State of the art in the delivery of photosensitizers for photodynamic therapy. *J. Photochem. Photobiol. B*, **66** (2), 89–106.

63 Wilson, B.C., Patterson, M.S., and Lilge, L. (1997) Implicit and explicit dosimetry in photodynamic therapy. *Lasers Med. Sci.*, **12**, 182–199.

64 van Veen, R.L., Robinson, D.J., Siersema, P.D., and Sterenborg, H.J. (2006) The importance of in situ dosimetry during photodynamic therapy of Barrett's esophagus. *Gastrointest. Endosc.*, **64** (5), 786–788.

65 Zhou, X., Pogue, B.W., Chen, B., Demidenko, E., Joshi, R., Hoopes, J., and Hasan, T. (2006) Pretreatment photosensitizer dosimetry reduces variation in tumor response. *Int. J. Radiat. Oncol. Biol. Phys.*, **64** (4), 1211–1220.

66 Star, W.M. (1997) Light dosimetry in vivo. *Phys. Med. Biol.*, **42** (5), 763–787.

67 Braichotte, D.R., Savary, J.F., Monnier, P., and van den Bergh, H.E. (1996) Optimizing light dosimetry in photodynamic therapy of early stage carcinomas of the esophagus using fluorescence spectroscopy. *Lasers Surg. Med.*, **19** (3), 340–346.

68 Farrell, T.J., Wilson, B.C., Patterson, M.S., and Olivo, M.C. (1998) Comparison of the in vivo photodynamic threshold dose for photofrin, mono- and tetrasulfonated aluminum phthalocyanine using a rat liver model. *Photochem. Photobiol.*, **68** (3), 394–399.

69 Wang, H.W., Rickter, E., Yuan, M., Wileyto, E.P., Glatstein, E., Yodh, A., and Busch, T.M. (2007) Effect of photosensitizer dose on fluence rate responses to photodynamic therapy. *Photochem. Photobiol.*, **83** (5), 1040–1048.

70 Yuan, J., Mahama-Relue, P.A., Fournier, R.L., and Hampton, J.A. (1997) Predictions of mathematical models of tissue oxygenation and generation of singlet oxygen during photodynamic therapy. *Radiat. Res.*, **148** (4), 386–394.

71 Georgakoudi, I. and Foster, T.H. (1998) Singlet oxygen- versus nonsinglet oxygen-mediated mechanisms of sensitizer photobleaching and their effects on photodynamic dosimetry. *Photochem. Photobiol.*, **67** (6), 612–625.

72 Mitra, S. and Foster, T.H. (2005) Photophysical parameters, photosensitizer retention and tissue optical properties completely account for the higher photodynamic efficacy of meso-tetra-hydroxyphenyl-chlorin vs photofrin. *Photochem. Photobiol.*, **81** (4), 849–859.

73 Finlay, J.C., Mitra, S., Patterson, M.S., and Foster, T.H. (2004) Photobleaching kinetics of photofrin in vivo and in multicell tumour spheroids indicate two simultaneous bleaching mechanisms. *Phys. Med. Biol.*, **49** (21), 4837–4860.

74 Wang, K.K., Mitra, S., and Foster, T.H. (2007) A comprehensive mathematical model of microscopic dose deposition in photodynamic therapy. *Med. Phys.*, **34** (1), 282–293.

75 Wang, K.K., Mitra, S., and Foster, T.H. (2008) Photodynamic dose does not correlate with long-term tumor response to mthpc-pdt performed at several drug-light intervals. *Med. Phys.*, **35** (8), 3518–3526.

76 Seshadri, M., Bellnier, D.A., Vaughan, L.A., Spernyak, J.A., Mazurchuk, R., Foster, T.H., and Henderson, B.W. (2008) Light delivery over extended time periods enhances the effectiveness of photodynamic therapy. *Clin. Cancer Res.*, **14** (9), 2796–2805.

77 Woodhams, J.H., MacRobert, A.J., and Bown, S.G. (2007) The role of oxygen monitoring during photodynamic therapy and its potential for treatment dosimetry. *Photochem. Photobiol. Sci.*, **6** (12), 1246–1256.

78 Pogue, B.W. and Hasan, T. (1997) A theoretical study of light fractionation and dose-rate effects in photodynamic therapy. *Radiat. Res*, **147** (5), 551–559.

79 Boere, I.A., Robinson, D.J., de Bruijn, H.S., Kluin, J., Tilanus, H.W., Sterenborg, H.J., and de Bruin, R.W. (2006) Protoporphyrin IX fluorescence photobleaching and the response of rat Barrett's esophagus following 5-aminolevulinic acid photodynamic therapy. *Photochem. Photobiol.*, **82** (6), 1638–1644.

80 Cottrell, W.J., Paquette, A.D., Keymel, K.R., Foster, T.H., and Oseroff, A.R. (2008) Irradiance-dependent photobleaching and pain in delta-aminolevulinic acid-photodynamic therapy of superficial basal cell carcinomas. *Clin. Cancer Res.*, **14** (14), 4475–4483.

81 Dysart, J.S., Singh, G., and Patterson, M.S. (2005) Calculation of singlet oxygen dose from photosensitizer fluorescence and photobleaching during mthpc photodynamic therapy of mll cells. *Photochem. Photobiol.*, **81** (1), 196–205.

82 Ericson, M.B., Sandberg, C., Stenquist, B., Gudmundson, F., Karlsson, M., Ros, A.M., Rosen, A., Larko, O., Wennberg, A.M., and Rosdahl, I. (2004) Photodynamic therapy of actinic keratosis at varying fluence rates: assessment of photobleaching, pain and primary clinical outcome. *Br. J. Dermatol.*, **151** (6), 1204–1212.

83 Sheng, C., Hoopes, P.J., Hasan, T., and Pogue, B.W. (2007) Photobleaching-based dosimetry predicts deposited dose in ala-ppix pdt of rodent esophagus. *Photochem. Photobiol.*, **83** (3), 738–748.

84 Zeng, H., Korbelik, M., McLean, D.I., MacAulay, C., and Lui, H. (2002) Monitoring photoproduct formation and photobleaching by fluorescence spectroscopy has the potential to improve pdt dosimetry with a verteporfin-like photosensitizer. *Photochem. Photobiol.*, **75** (4), 398–405.

85 Farrell, T.J., Hawkes, R.P., Patterson, M.S., and Wilson, B.C. (1998) Modeling of photosensitizer fluorescence emission and photobleaching for photodynamic therapy dosimetry. *Appl. Opt.*, **37** (31), 7168–7183.

86 Potter, W.R., Mang, T.S., and Dougherty, T.J. (1987) The theory of photodynamic therapy dosimetry: consequences of photo-destruction of sensitizer. *Photochem. Photobiol.*, **46** (1), 97–101.

87 Robinson, D.J., de Bruijn, H.S., van der Veen, N., Stringer, M.R., Brown, S.B., and Star, W.M. (1998) Fluorescence photobleaching of ala-induced protoporphyrin IX during photodynamic therapy of normal hairless mouse skin: the effect of light dose and irradiance and the resulting biological effect. *Photochem. Photobiol.*, **67** (1), 140–149.

88 Kruijt, B., de Bruijn, H.S., van der Ploeg-van den Heuvel, A., de Bruin, R.W., Sterenborg, H.J., Amelink, A., and Robinson, D.J. (2008) Monitoring ala-induced ppix photodynamic therapy in the rat esophagus using fluorescence and reflectance spectroscopy. *Photochem. Photobiol.*, **84** (6), 1515–1527.

89 Jongen, A.J. and Sterenborg, H.J. (1997) Mathematical description of photobleaching in vivo describing the influence of tissue optics on measured fluorescence signals. *Phys. Med. Biol.*, **42** (9), 1701–1716.

90 Jarvi, M.T., Niedre, M.J., Patterson, M.S., and Wilson, B.C. (2006) Singlet oxygen luminescence dosimetry (sold) for photodynamic therapy: current status, challenges and future prospects. *Photochem. Photobiol.*, **82** (5), 1198–1210.

91 Yamamoto, J., Yamamoto, S., Hirano, T., Li, S., Koide, M., Kohno, E., Okada, M., Inenaga, C., Tokuyama, T., Yokota, N.,

Terakawa, S., and Namba, H. (2006) Monitoring of singlet oxygen is useful for predicting the photodynamic effects in the treatment for experimental glioma. *Clin. Cancer Res.*, **12** (23), 7132–7139.

92 Niedre, M.J., Secord, A.J., Patterson, M.S., and Wilson, B.C. (2003) *In vitro* tests of the validity of singlet oxygen luminescence measurements as a dose metric in photodynamic therapy. *Cancer Res.*, **63** (22), 7986–7994.

93 Hatz, S., Poulsen, L., and Ogilby, P.R. (2008) Time-resolved singlet oxygen phosphorescence measurements from photosensitized experiments in single cells: effects of oxygen diffusion and oxygen concentration. *Photochem. Photobiol.*, **84** (5), 1284–1290.

94 Yu, G., Durduran, T., Zhou, C., Zhu, T.C., Finlay, J.C., Busch, T.M., Malkowicz, S.B., Hahn, S.M., and Yodh, A.G. (2006) Real-time in situ monitoring of human prostate photodynamic therapy with diffuse light. *Photochem. Photobiol.*, **82** (5), 1279–1284.

95 Enejder, A.M., Klinteberg, C.af., Wang, I., Andersson-Engels, S., Bendsoe, N., Svanberg, S., and Svanberg, K. (2000) Blood perfusion studies on basal cell carcinomas in conjunction with photodynamic therapy and cryotherapy employing laser-doppler perfusion imaging. *Acta Derm. Venereol.*, **80** (1), 19–23.

96 Standish, B.A., Lee, K.K., Jin, X., Mariampillai, A., Munce, N.R., Wood, M.F., Wilson, B.C., Vitkin, I.A., and Yang, V.X. (2008) Interstitial doppler optical coherence tomography as a local tumor necrosis predictor in photodynamic therapy of prostatic carcinoma: an *in vivo* study. *Cancer Res.*, **68** (23), 9987–9995.

97 Yu, G., Durduran, T., Zhou, C., Wang, H.W., Putt, M.E., Saunders, H.M., Sehgal, C.M., Glatstein, E., Yodh, A.G., and Busch, T.M. (2005) Noninvasive monitoring of murine tumor blood flow during and after photodynamic therapy provides early assessment of therapeutic efficacy. *Clin. Cancer Res.*, **11** (9), 3543–3552.

98 Blaha, G.R. III, Wertz, F.D., and Marx, J.L. (2008) Profound choroidal hypoperfusion after combined photodynamic therapy and intravitreal triamcinolone acetonide. *Ophthalm. Surg. Lasers Imag.*, **39** (1), 6–11.

99 Gross, S., Gilead, A., Scherz, A., Neeman, M., and Salomon, Y. (2003) Monitoring photodynamic therapy of solid tumors online by bold-contrast MRI. *Nat. Med.*, **9** (10), 1327–1331.

100 Gersing, E., Kelleher, D.K., and Vaupel, P. (2003) Tumour tissue monitoring during photodynamic and hyperthermic treatment using bioimpedance spectroscopy. *Physiol. Meas.*, **24** (2), 625–637.

101 Banihashemi, B., Vlad, R., Debeljevic, B., Giles, A., Kolios, M.C., and Czarnota, G.J. (2008) Ultrasound imaging of apoptosis in tumor response: novel preclinical monitoring of photodynamic therapy effects. *Cancer Res.*, **68** (20), 8590–8596.

102 Berard, V., Lecomte, R., and Van Lier, J.E. (2006) Positron emission tomography imaging of tumor response after photodynamic therapy. *J. Environ. Pathol. Toxicol. Oncol.*, **25** (1–2), 239–249.

103 Plaks, V., Koudinova, N., Nevo, U., Pinthus, J.H., Kanety, H., Eshhar, Z., Ramon, J., Scherz, A., Neeman, M., and Salomon, Y. (2004) Photodynamic therapy of established prostatic adenocarcinoma with tookad: a biphasic apparent diffusion coefficient change as potential early MRI response marker. *Neoplasia*, **6** (3), 224–233.

104 Moriyama, E.H., Bisland, S.K., Lilge, L., and Wilson, B.C. (2004) Bioluminescence imaging of the response of rat gliosarcoma to ala-ppix-mediated photodynamic therapy. *Photochem. Photobiol.*, **80** (2), 242–249.

105 Augustin, A.J., Puls, S., and Offermann, I. (2007) Triple therapy for choroidal neovascularization due to age-related macular degeneration: verteporfin pdt, bevacizumab, and dexamethasone. *Retina*, **27** (2), 133–140.

106 Michels, S., Hansmann, F., Geitzenauer, W., and Schmidt-Erfurth, U. (2006) Influence of treatment parameters on selectivity of verteporfin therapy. *Invest. Ophthalmol. Vis. Sci.*, **47** (1), 371–376.

107 Samkoe, K.S., Clancy, A.A., Karotki, A., Wilson, B.C., and Cramb, D.T. (2007) Complete blood vessel occlusion in the chick chorioallantoic membrane using two-photon excitation photodynamic therapy: implications for treatment of wet age-related macular degeneration. *J. Biomed. Opt.*, **12** (3), 034025.

108 Overholt, B.F., Lightdale, C.J., Wang, K.K., Canto, M.I., Burdick, S., Haggitt, R.C., Bronner, M.P., Taylor, S.L., Grace, M.G., and Depot, M. (2005) Photodynamic therapy with porfimer sodium for ablation of high-grade dysplasia in Barrett's esophagus: international, partially blinded, randomized phase iii trial. *Gastrointest. Endosc.*, **62** (4), 488–498.

109 Triesscheijn, M., Baas, P., Schellens, J.H., and Stewart, F.A. (2006) Photodynamic therapy in oncology. *Oncologist.*, **11** (9), 1034–1044.

110 Biel, M. (2006) Advances in photodynamic therapy for the treatment of head and neck cancers. *Lasers Surg. Med.*, **38** (5), 349–355.

111 Jerjes, W., Upile, T., Betz, C.S., El Maaytah, M., Abbas, S., Wright, A., and Hopper, C. (2007) The application of photodynamic therapy in the head and neck. *Dent. Update*, **34** (8), 478–486.

112 Ortner, M.A. and Dorta, G. (2006) Technology insight: photodynamic therapy for cholangiocarcinoma. *Nat. Clin. Pract. Gastroenterol. Hepatol.*, **3** (8), 459–467.

113 Fan, K.F., Hopper, C., Speight, P.M., Buonaccorsi, G.A., and Bown, S.G. (1997) Photodynamic therapy using mthpc for malignant disease in the oral cavity. *Int. J. Cancer*, **73** (1), 25–32.

114 Hetzel, F.W., Brahmavar, S.M., Chen, Q., Jacques, S.L., Patterson, M.S., Wilson, B.C., and Zhu, T.C. (2005) Photodynamic Therapy Dosimetry: a Task Group Report of the General Medical Physics Committee, AAPM Report No. 88, Published by Medical Physics Publishing for American Association of Physicists in Medicine.

115 Nyst, H.J., van Veen, R.L., Tan, I.B., Peters, R., Spaniol, S., Robinson, D.J., Stewart, F.A., Levendag, P.C., and Sterenborg, H.J. (2007) Performance of a dedicated light delivery and dosimetry device for photodynamic therapy of nasopharyngeal carcinoma: phantom and volunteer experiments. *Lasers Surg. Med*, **39** (8), 647–653.

116 Wang, H.W., Zhu, T.C., Putt, M.E., Solonenko, M., Metz, J., Dimofte, A., Miles, J., Fraker, D.L., Glatstein, E., Hahn, S.M., and Yodh, A.G. (2005) Broadband reflectance measurements of light penetration, blood oxygenation, hemoglobin concentration, and drug concentration in human intraperitoneal tissues before and after photodynamic therapy. *J. Biomed. Opt.*, **10** (1), 14004.

117 Star, W.M., Aalders, M.C., Sac, A., and Sterenborg, H.J. (2002) Quantitative model calculation of the time-dependent protoporphyrin IX concentration in normal human epidermis after delivery of ala by passive topical application or lontophoresis. *Photochem. Photobiol.*, **75** (4), 424–432.

118 Svaasand, L.O., Wyss, P., Wyss, M.T., Tadir, Y., Tromberg, B.J., and Berns, M.W. (1996) Dosimetry model for photo-dynamic therapy with topically administered photosensitizers. *Lasers Surg. Med.*, **18** (2), 139–149.

119 Klinteberg, C.af., Enejder, A.M., Wang, I., Andersson-Engels, S., Svanberg, S., and Svanberg, K. (1999) Kinetic fluorescence studies of 5-aminolaevulinic acid-induced protoporphyrin IX accumulation in basal cell carcinomas. *J. Photochem. Photobiol. B*, **49** (2–3), 120–128.

120 Beck, T.J., Kreth, F.W., Beyer, W., Mehrkens, J.H., Obermeier, A., Stepp, H., Stummer, W., and Baumgartner, R. (2007) Interstitial photodynamic therapy of nonresectable malignant glioma recurrences using 5-aminolevulinic acid induced protoporphyrin IX. *Lasers Surg. Med.*, **39** (5), 386–393.

121 Waidelich, R., Beyer, W., Knuchel, R., Stepp, H., Baumgartner, R., Schroder, J., Hofstetter, A., and Kriegmair, M. (2003) Whole bladder photodynamic therapy

with 5-aminolevulinic acid using a white light source. *Urology*, **61** (2), 332–337.

122 Laubach, H.J., Chang, S.K., Lee, S., Rizvi, I., Zurakowski, D., Davis, S.J., Taylor, C.R., and Hasan, T. (2008) *In-vivo* singlet oxygen dosimetry of clinical 5-aminolevulinic acid photodynamic therapy. *J. Biomed. Opt.*, **13** (5), 050504.

123 Johansson, A., Axelsson, J., Andersson-Engels, S., and Swartling, J. (2007) Realtime light dosimetry software tools for interstitial photodynamic therapy of the human prostate. *Med. Phys.*, **34** (11), 4309–4321.

124 Weersink, R.A., Bogaards, A., Gertner, M., Davidson, S.R., Zhang, K., Netchev, G., Trachtenberg, J., and Wilson, B.C. (2005) Techniques for delivery and monitoring of tookad (wst09)-mediated photodynamic therapy of the prostate: clinical experience and practicalities. *J. Photochem. Photobiol. B*, **79** (3), 211–222.

125 Zhu, T.C., Finlay, J.C., and Hahn, S.M. (2005) Determination of the distribution of light, optical properties, drug concentration, and tissue oxygenation in-vivo in human prostate during motexafin lutetium-mediated photodynamic therapy. *J. Photochem. Photobiol. B*, **79** (3), 231–241.

126 Vogl, T.J., Eichler, K., Mack, M.G., Zangos, S., Herzog, C., Thalhammer, A., and Engelmann, K. (2004) Interstitial photodynamic laser therapy in interventional oncology. *Eur. Radiol.*, **14** (6), 1063–1073.

127 Johansson, A., Johansson, T., Thompson, M.S., Bendsoe, N., Svanberg, K., Svanberg, S., and Andersson-Engels, S. (2006) *In vivo* measurement of parameters of dosimetric importance during interstitial photodynamic therapy of thick skin tumors. *J. Biomed. Opt.*, **11** (3), 34029.

128 Betz, C.S., Jager, H.R., Brookes, J.A., Richards, R., Leunig, A., and Hopper, C. (2007) Interstitial photodynamic therapy for a symptom-targeted treatment of complex vascular malformations in the head and neck region. *Lasers Surg. Med.*, **39** (7), 571–582.

129 Li, J. and Zhu, T.C. (2008) Determination of *in vivo* light fluence distribution in a heterogeneous prostate during photodynamic therapy. *Phys. Med. Biol.*, **53** (8), 2103–2114.

130 Jankun, J., Keck, R.W., Skrzypczak-Jankun, E., Lilge, L., and Selman, S.H. (2005) Diverse optical characteristic of the prostate and light delivery system: implications for computer modelling of prostatic photodynamic therapy. *BJU. Int.*, **95** (9), 1237–1244.

131 Axelsson, J., Swartling, J., and Andersson-Engels, S. (2009) *In vivo* photosensitizer tomography inside the human prostate. *Opt. Lett.*, **34** (3), 232–234.

132 Palsson, S., Gustafsson, L., Bendsoe, N., Soto, T.M., Andersson-Engels, S., and Svanberg, K. (2003) Kinetics of the superficial perfusion and temperature in connection with photodynamic therapy of basal cell carcinomas using esterified and non-esterified 5-aminolaevulinic acid. *Br. J. Dermatol.*, **148** (6), 1179–1188.

133 Juzeniene, A., Nielsen, K.P., Zhao, L., Ryzhikov, G.A., Biryulina, M.S., Stamnes, J.J., Stamnes, K., and Moan, J. (2008) Changes in human skin after topical pdt with hexyl aminolevulinate. *Photodiag. Photodyn. Ther.*, **5** (3), 176–181.

134 Kruijt, B., de Bruijn, H.S., van der Ploeg-van den Heuvel, A., Sterenborg, H.J., and Robinson, D.J. (2006) Laser speckle imaging of dynamic changes in flow during photodynamic therapy. *Lasers Med. Sci.*, **21** (4), 208–212.

135 Mennel, S., Barbazetto, I., Meyer, C.H., Peter, S., and Stur, M. (2007) Ocular photodynamic therapy - standard applications and new indications (part 1). *Ophthalmologica*, **221** (4), 216–226.

136 Mennel, S., Barbazetto, I., Meyer, C.H., Peter, S., and Stur, M. (2007) Ocular photodynamic therapy – standard applications and new indications (part 2). *Ophthalmologica*, **221** (5), 282–291.

137 Miyazawa, S., Nishida, K., Komiyama, T., Nakae, Y., Takeda, K., Yorimitsu, M., Kitamura, A., Kunisada, T., Ohtsuka, A., and Inoue, H. (2006) Novel transdermal photodynamic therapy using atx-s10.na(ii) induces apoptosis of

synovial fibroblasts and ameliorates collagen antibody-induced arthritis in mice. *Rheumatol. Int.*, **26** (8), 717–725.

138 Trauner, K.B., Gandour-Edwards, R., Bamberg, M., Shortkroff, S., Sledge, C., and Hasan, T. (1998) Photodynamic synovectomy using benzoporphyrin derivative in an antigen-induced arthritis model for rheumatoid arthritis. *Photochem. Photobiol.*, **67** (1), 133–139.

139 Maisch, T. (2007) Anti-microbial photodynamic therapy: useful in the future? *Lasers Med. Sci.*, **22** (2), 83–91.

140 Berg, K., Dietze, A., Kaalhus, O., and Høgset, A. (2005) Site-specific drug delivery by photochemical internalization enhances the antitumor effect of bleomycin. *Clin. Cancer Res.*, **11** (23), 8476–8485.

141 Berg, K., Folini, M., Prasmickaite, L., Selbo, P.K., Bonsted, A., Engesaeter, B., Zaffaroni, N., Weyergang, A., Dietze, A., Maelandsmo, G.M., Wagner, E., Norum, O.-J, and Høgset, A. (2007) Photochemical internalization: a new tool for drug delivery. *Curr. Pharm. Biotechnol.*, **8** (6), 362–372.

142 Hirschberg, H., Zhang, M.J., Gach, H.M., Uzal, F.A., Peng, Q., Sun, C.H., Chighvinadze, D., and Madsen, S.J. (2009) Targeted delivery of bleomycin to the brain using photo-chemical internalization of clostridium perfringens epsilon prototoxin. *J. Neurooncol..* **95** (3), 317–329.

143 Biolo, R., Jori, G., Kennedy, J.C., Nadeau, P., Pottier, R., Reddi, E., and Weagle, G. (1991) A comparison of fluorescence methods used in the pharmacokinetic studies of Zn(II) phthalocyanine in mice. *Photochem. Photobiol.*, **53** (1), 113–118.

144 Cannon, J.B. (1993) Pharmaceutics and drug delivery aspects of heme and porphyrin therapy. *J. Pharm. Sci.*, **82** (5), 435–446.

145 Pegaz, B., Debefve, E., Ballini, J.-P., Wagnières, G., Spaniol, S., Albrecht, V., Scheglmann, D.V., Nifantiev, N.E., van den Bergh, H., and Konan-Kouakou, Y.N. (2006) Photothrombic activity of m-thpc-loaded liposomal formulations: pre-clinical assessment on chick chorioallantoic membrane model. *Eur. J. Pharm. Sci.*, **28** (1–2), 134–140.

146 van Dongen, G.A.M.S., Visser, G.W.M., and Vrouenraets, M.B. (2004) Photosensitizer-antibody conjugates for detection and therapy of cancer. *Adv. Drug. Deliv. Rev.*, **56** (1), 31–52.

147 Soukos, N.S., Hamblin, M.R., Keel, S., Fabian, R.L., Deutsch, T.F., and Hasan, T. (2001) Epidermal growth factor receptor-targeted immunophotodiagnosis and photoimmunotherapy of oral precancer *in vivo. Cancer Res.*, **61** (11), 4490–4496.

148 Chen, J., Stefflova, K., Niedre, M.J., Wilson, B.C., Chance, B., Glickson, J.D., and Zheng, G. (2004) Protease-triggered photosensitizing beacon based on singlet oxygen quenching and activation. *J. Am. Chem. Soc.*, **126** (37), 11450–11451.

149 Lo, P.-C., Chen, J., Stefflova, K., Warren, M.S., Navab, R., Bandarchi, B., Mullins, S., Tsao, M., Cheng, J.D., and Zheng, G. (2009) Photodynamic molecular beacon triggered by fibroblast activation protein on cancer-associated fibroblasts for diagnosis and treatment of epithelial cancers. *J. Med. Chem.*, **52** (2), 358–368.

150 Zheng, G., Chen, J., Stefflova, K., Jarvi, M., Li, H., and Wilson, B.C. (2007) Photodynamic molecular beacon as an activatable photosensitizer based on protease-controlled singlet oxygen quenching and activation. *Proc. Natl. Acad. Sci. USA*, **104** (21), 8989–8994.

151 Wilson, B.C. (2006) Photonic and non-photonic based nanoparticles in cancer imaging and therapeutics, in *Photon-based Nanoscience and Nanobiotechnology* (eds J.J. Dubowski and S. Tanev), NATO Science Series, Vol. 239, Springer, pp. 121–157.

152 Kovalev, D. and Fujii, M. (2005) Silicon nanocrystals: photosensitizers for oxygen molecules. *Adv. Mater.*, **17** (21), 2531–2544.

153 Samia, A.C.S., Dayal, S., and Burda, C. (2006) Quantum dot-based energy transfer: perspectives and potential for applications in photodynamic therapy. *Photochem. Photobiol.*, **82** (3), 617–625.

154 Luksiene, Z., Juzenas, P., and Moan, J. (2006) Radiosensitization of tumours by porphyrins. *Cancer Lett.*, **235** (1), 40–47.

155 Juzenas, P., Chen, W., Sun, Y.-P., Coelho, M.A.N., Generalov, R., Generalova, N., and Christensen, I.L. (2008) Quantum dots and nanoparticles for photodynamic and radiation therapies of cancer. *Adv. Drug Deliv. Rev.*, **60** (15), 1600–1614.

156 Johansson, T., Thompson, M.S., Stenberg, M., Klinteberg, C.af., Stefan, A.-E., Sune, S., and Svanberg, K. (2002) Feasibility study of a system for combined light dosimetry and interstitial photodynamic treatment of massive tumors. *Appl. Opt.*, **41** (7), 1462–1468.

157 Soto-Thompson, M., Johansson, A., Johansson, T., Andersson-Engels, S., Svanberg, S., Bendsoe, N., and Svanberg, K. (2005) Clinical system for interstitial photodynamic therapy with combined on-line dosimetry measurements. *Appl. Opt.*, **44** (19), 4023–4031.

158 Thompson, M.S., Andersson-Engels, S., Svanberg, S., Johansson, T., Palsson, S., Bendsoe, N., Derjabo, A., Kapostins, J., Unne, S., Spigulis, J., and Svanberg, K. (2006) Photodynamic therapy of nodular basal cell carcinoma with multifiber contact light delivery. *J. Environ. Pathol. Toxicol. Oncol.*, **25** (1–2), 411–424.

9

Laser Welding of Biological Tissue: Mechanisms, Applications and Perspectives

Paolo Matteini, Francesca Rossi, Fulvio Ratto, and Roberto Pini

9.1
Introduction

Laser welding of biological tissues has been proposed in several surgical fields for the closure of wounds in controlled laboratory tests over the last 30 years. Joining tissues by application of laser irradiation was first reported at the end of the 1970s, when a neodymium:YAG laser was used successfully to join small blood vessels [1]. Since then, several experiments have been performed using various lasers for sealing many tissue types, including blood vessels, nerves, skin, urethra, stomach, and colon (see also previous reviews [2, 3]). Laser welding has progressively gained relevance in the clinical setting, where it now appears as a valid alternative to standard surgical techniques.

According to a general perspective, lasers hold the promise of providing instantaneous, watertight seals, which is important in many critical surgeries, such as, for example, for gastrointestinal and vascular repairs, without the introduction of foreign materials, such as sutures or staples. Nowadays, due to increasing demand for minimally invasive procedures, laser welding is emerging as the appropriate technique for applications in which suturing and stapling is particularly difficult, such as in microsurgery, laparoscopy or endoscopy, or for the treatment of extremely thin tissues. Other advantages over conventional suturing include reduced operation times, fewer skill requirements, decreased foreign-body reaction and therefore reduced inflammatory response, increased ability to induce regeneration of the original tissue architecture, and an improved cosmetic appearance. The final aim of this procedure is to improve the quality of life of patients, by reduction of healing times and the risk of postoperative complications.

The laser–tissue interaction occurring during a laser-mediated welding of biological tissues is unanimously considered to be photothermal [2, 3]. This interaction is distinguished by the absorption of photons emitted by the laser source, which generates heat through a target volume. The thermal changes induced within the tissue about the lesion result, in turn, in a bond between its adjoining edges. The heat is produced through the absorption of the laser energy by endogenous chromophores

Laser Imaging and Manipulation in Cell Biology. Edited by Francesco S. Pavone
Copyright © 2010 WILEY-VCH Verlag GmbH & Co. KGaA, Weinheim
ISBN: 978-3-527-40929-7

such as the water content of the tissue at infrared wavelengths and the hemoglobin and melanin at visible wavelengths.

Sometimes, photochemical welding of tissues has also been investigated as an alternative method for tissue repair without direct use of heat [4]. Photochemical interactions can be typically achieved at low energy levels that suffice to activate the chemical processes behind tissue closuring. This technique utilizes chemical cross-linking agents (such as riboflavin, fluorescein, etc.) applied to the cut, which upon light-activation produce covalent crosslinks between the structures of the native tissue.

Many types of lasers have been proposed for laser tissue welding. Infrared and near-infrared sources include carbon dioxide (CO_2), thulium-holmium-chromium, holmium, thulium, and neodymium rare-earth-doped-garnets (THC:YAG, Ho:YAG, Tm:YAG, and Nd:YAG, respectively), and gallium aluminium arsenide diode (GaAlAs) lasers. Visible sources include potassium-titanyl phosphate (KTP) frequency-doubled Nd:YAG, and argon lasers. Among the infrared sources previously proposed to induce the water absorption, and thus the wound closure, the CO_2 laser has long been the most popular. This laser has been especially useful for laser repairs of thin tissues because of its short penetration depth ($<20\,\mu m$). However, when dealing with thick tissues, welding can only be achieved by irradiating with higher laser powers and longer exposure times, and this frequently induces high levels of heat damage [5]. The emissions of other lasers, such as near-infrared neodymium:YAG and diode lasers, are more suited for the welding of thicker tissues. In all cases, control of the dosimetry of the laser irradiation and of the induced temperature rise is crucial to minimize the risk of irreversible damage to the tissue. In fact, since water, hemoglobin, and melanin are the main absorbers of laser light, the heating effect is not selectively limited to a target area, and all irradiated tissues are heated. If laser radiation is used in excess, an undesired thermal damage may be induced in the surrounding area, resulting in a poor strength of the weld. This is the main reason behind the disproportion between the large number of experimental studies reported in the literature and the limited number of clinical applications.

Two improvements have been useful in addressing these problems associated with laser tissue welding: the application of exogenous chromophores and the addition of biomaterials to be employed as solders.

To improve the localization of laser light absorption, the application of photo-enhancing dyes to the tissue has been proposed. These exogenous optical chromophores have enabled the fusion of wounds selectively and at lower irradiation fluences, thus reducing the risk of excessive thermal damage to surrounding tissues. In fact, acting as wavelength-specific absorbers, they make it possible to achieve differential absorption of the laser light between the stained region and the surroundings. In practice, while the usage of a wavelength strongly absorbed by the endogenous chromophores induces very high temperatures in the area under direct irradiation, the combination of a proper exogenous chromophore and a laser wavelength at which the tissue is almost transparent allows for a confined and homogenous temperature enhancement deep in the stained tissue. The resulting welding effect may be modulated in the depth of the tissue (Figure 9.1), thus resulting in a more effective closure of the wound. Various chromophores have been employed

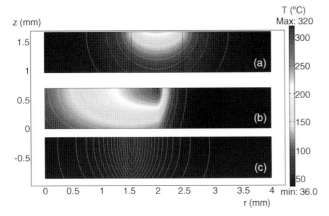

Figure 9.1 Induced temperature maps in a corneal tissue during three different laser welding procedures. This image depicts the results of the computational models of three different irradiation conditions of the implementation of the laser welding of the cornea: (a) a CO_2 laser light impinging on the external surface of a cornea, and the direction is parallel to the z axis [82]; (b) an erbium laser is scanned onto the external surface of the cornea sample, having light propagation in the z axis direction and a constant linear velocity parallel to the r axis direction, from left to right [24]; (c) an 810 nm diode laser is absorbed by an ICG-stained cornea area and the laser light is delivered in the side irradiation procedure (propagation axis parallel to the horizontal axis, from left to right) [8]. In (a) and (b) the laser light is absorbed by the water content of the tissue, while in (c) it is absorbed by the ICG stained wound, producing a much more homogeneous distribution of heat in depth. The model was solved by the finite element method (Comsol Multiphysics 3.5a, Comsol AB, Sweden). In all three cases a 2-s irradiation time was considered. These results are only an estimation of the induced temperature range, because the dependence of the opto-thermal parameters on temperature (T) was not considered.

as photo-enhancing dyes, including indocyanine green (ICG), fluorescein, basic fuchsin, and India ink [6, 7]. A very popular setting of tissue laser welding includes the use of a near-infrared laser, which is poorly absorbed by the biological tissue, in conjunction with the topical application of a chromophore absorbing in the same spectral region. Current examples of this modality are in the transplant of the cornea, in cataract surgery, vascular tissue welding, skin welding, and laryngotracheal mucosa transplant [6, 8, 9]. In all these cases diode lasers emitting around 800 nm and the topical application of ICG have been used.

The addition of solders has been proposed to strengthen the seal both during and after laser application. Laser welding by means of solders, which are usually biological in nature, is termed "laser soldering". Following laser irradiation, photo-activation of the solder occurs, which makes it act as a glue that can form an interdigitated matrix among the collagen fibers. Usually, solders provide a large surface area over which fusion with the tissue can occur. This favors the approximation of the wound edges, which are well-arranged to heal together in the postoperative period (with deposition of extracellular matrix material). The most common procedures makes use of a topical protein preparation such as albumin and fibrinogen [7, 9, 10]. Numerous studies have confirmed the use of albumin as a suitable agent to enhance the laser weld [11, 12]. The great interest in the exploitation of human albumin as a solder for laser tissue welding is based on several advantages. It is easy to prepare and can be stored in solution without refrigeration for a long time. Moreover, the FDA-approved human albumin, which is commercially available, grants high levels of safety thanks to an effective inactivation of viruses during preparation.

To confine the heat into the area of solder application, incorporation into the solder of an exogenous chromophore was proposed, which has been shown to reduce the extent of collateral heat damage to adjacent tissues [10]. Effective results using chromophore-doped solders have been reported for various tissues as arteries and skin. Another improvement of traditional protein solders relies on the addition to the albumin solution of biopolymers, such as PLGA (poly(DL-lactic-*co*-glycolic acid)), which act as porous scaffolds [13]. These provide better flexibility as well as improved repair strength over albumin solders alone. Moreover, these materials can be loaded with a wide range of drugs and additives as antibiotics, anesthetics, and various growing factors, which provide a faster and better wound healing process [14]. The scaffolds are easily degraded *in vivo* and are eliminated through normal metabolic pathways. More recently nonproteic materials were proposed for laser soldering such as deacetylated chitosan [15]. Chitosan is a biocompatible and biodegradable polysaccharide extracted from the shell of crustaceans. In this case the unique mimetic properties of this polysaccharide are exploited to favor an optimal restructuring of the extracellular matrix during the healing phase. Moreover, of utility for tissue repair is its proven antimicrobial properties, which reduce the potential for infections.

9.2
Mechanism of Thermal Laser Welding

Understanding the mechanism responsible for the sealing process mediated by laser light is important for the appropriate selection of laser parameters (power density, pulse duration, and spot size) for the tissue to be welded. In this way, a combination of maximal bonding strength and minimal thermal damage can be achieved. According to present understanding and interpretation, laser welding is a soft thermal treatment of wound edges in which laser radiation is absorbed by the tissue, either directly from its principal absorbers such as water or from exogenous chromophores applied topically to the wound area. In the heated tissue the main components undergo thermal modifications, resulting in fusion of the apposed flaps.

9.2.1
Composition of the Extracellular Matrix

The main components of the extracellular matrix of biological tissues include proteins, mainly collagen, and nonproteic material, which is principally constituted of proteoglycans and glycosaminoglycans.

Collagen is the most abundant protein in the body and the main component of connective tissues. This protein has a triple helix composed of polypeptide chains highly stabilized by interconnecting hydrogen bonds. Within the collagen family, the main component of fibril-forming collagens is type I collagen, which provides the structural and organizational framework for skin, blood vessels, bone, tendon, cornea, and other tissues. In this type of collagen, molecules are packed in a quarter-staggered manner and connected by covalent bonds to form microfibrils a few nanometers in diameter, which then combine to form collagen fibrils a few to a hundred nanometers in size. The regular arrangement of the collagen molecules gives rise to a structural periodicity along the microfibrils and fibrils with a repeated distance of 65–67 nm. This periodicity results in a cross striations pattern that is apparent under electron microscopy. The absence of this pattern indicates a loss in the spatial organization within the collagen fibrils and is often used as a marker for the onset of thermal denaturation. Some tissues are characterized by two to three hierarchical levels of organization at progressively larger scales, beyond the fibrils. For example, in tendon, fibrils organize in 1–10 μm thick fibers that in turn form bundles of fibers called fascicles. In corneal stroma, collagen fibrils have uniform diameters and self-organize with each other to form 1 to 2 μm sheets known as lamellae.

In connective tissues, collagen fibrils and interfibrillar proteoglycans (PGs) form a dense and well-organized mutually interconnecting network. The interfibrillar PGs usually display a globular protein part (head) to which one glycosaminoglycan (GAG) is attached (tail). The interfibrillar PGs constitute a well-defined, complex molecular chain system providing rigid bridges between the fibrils and thereby being responsible for the maintenance of the regular array of collagen fibrils. PGs are equidistant and orthogonally attached at specific sites of the collagen fibrils by their protein cores. The interaction of the PGs with the collagen fibrils is thought to be noncovalent and is characterized by a high affinity constant. By using a specific proteoglycan staining for electron microscopy, GAGs have been shown to form antiparallel doublets (dimers) that make it possible to maintain the relative positions among the adjacent collagen fibrils [16]. According to Scott's hypothesis, two GAG chains, one from each PG, form duplexes, covering the space between fibrils, anchored by protein cores attached to each fibril [16, 17]. In a recent paper a multifaceted analysis has brought to light that the model GAG, hyaluronan, can generate stable aggregates (polymer networks) in physiological solution [18]. These structures can be dissolved reversibly upon heating at temperatures exceeding 40 °C. This led to the hypothesis of the formation of junction zones mediated by weak heat-labile interactions, partly confirming Scott's hypothesis.

9.2.2
Thermal Modifications of Connective Tissues and Mechanism of Welding

When a connective tissue is subject to a temperature rise, its extracellular components undergo several morphological changes (Figure 9.2). The first step is governed by the thermal denaturation of PGs, mainly occurring before the onset of fibrillar collagen denaturation, that is, from ∼45 to ∼60 °C. In this range one or two subsequent phase transitions are observed that are correlated to a local disorder of

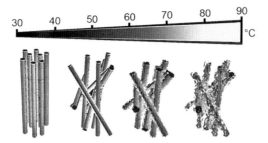

Figure 9.2 Thermal transitions of connective tissue from a collagen point of view. Native tissue is characterized by a regular arrangement of collagen fibrils. The first step leads to a progressive fibrillar disorganization, ascribed to the breaking of the proteoglycan bridges connecting collagen fibrils together. At higher temperatures fibrillar collagen is denatured during two subsequent steps: in the first, intramolecular H-bonds break, leading to a moderate thickening of the fibrillar size (followed by a shortening in length) and to the appearance of frayed fibrillar edges; in the second step, covalent crosslinks connecting collagen molecules together break, causing the complete denaturation of collagen (hyalinosis).

the regular fibrillar collagen arrangement. This can probably be ascribed to the breaking of interfibrillar PGs bridges, which leads to a drastic change of the interfibrillar distances [18, 19]. The next stage consists in the helix-to-coil transition of collagen molecules, which proceeds in two following steps. At first, unwinding of the triple helices occurs due to hydrolysis of the intramolecular hydrogen bonds. This leads to tissue shortening and to a loss of periodicity due to a shrinkage effect parallel to the axis of the fibrils [20]. As a consequence, fibrillar edges appear frayed and the average fibrillar diameter increases. At higher temperatures covalent crosslinks connecting collagen strands break, resulting in complete destruction of the fibrillar structure and causing full denaturation of collagen and relaxation of the tissue [21]. The onset of collagen shrinkage often quoted in the clinical literature is around 60 °C, while relaxation is reported to occur beyond 75 °C [22]. However, it is worth noting that collagen denaturation is a rate process governed by the local temperature/time response: thus the time/temperature profile influences the shrinkage threshold of collagen, as well as the relaxation phase [23]. Moreover, both the temperature for maximum shrinkage and the relaxation temperature depend on the crosslinks density, which rises on tissue aging as well as on hydroxyproline (one of the main amino acids constituting the collagen backbone) content, which varies among species and tissues.

The modifications occurring to collagen and proteoglycans upon photothermal laser treatment have been monitored mainly by means of different microscopy techniques (Figure 9.3). These include traditional methods such as optical and fluorescence microscopy, which allow for an investigation at the micron scale [6, 24], and transmission and scanning electron microscopy (TEM and SEM, respectively), which are useful when studying nanometric structures [19, 25, 26]. Other techniques recently applied for the analysis of biological material have also been tested, such as atomic force microscopy (AFM) and second-harmonic generation (SHG) microscopy, which provide complementary information [27–29]. These studies have been

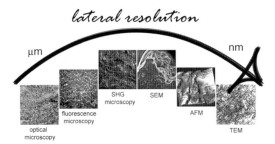

Figure 9.3 Main microscopic techniques employed for the study of the mechanism of laser tissue welding.

helpful (although not exhaustive) in elucidating the different dynamics behind the sealing process.

According to the scientific literature, different schemes can be invoked for the laser welding of biological tissue (Figure 9.4). At first, by considering the effects induced by the photothermal treatment, we can distinguish "hard," "moderate," and "soft" laser tissue welding. In other words, on the basis of the dosimetry of the laser irradiation, the choice of laser and the parameters setting, the tissue type, and the nature of the absorber (endogenous or exogenous), different dynamics of the extracellular matrix develop at the weld site. These can be typically distinguished by: (i) the photocoagulation of the collagen, (ii) "interdigitation" of collagen fibrils, and (iii) reorganization of the nonfibrillar components, which are associated with the "hard," "moderate," and "soft" laser welding, respectively.

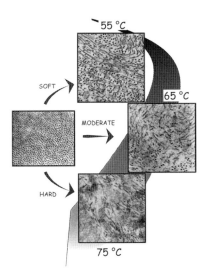

Figure 9.4 Transmission electron microscopy images of connective tissue showing the effects of "soft," "moderate," and "hard" laser welding as a function of the temperature induced at the weld site. A control sample showing the regular fibrillar arrangement is reported on the left-hand side.

9.2.2.1 Hard Laser Welding

Some authors have observed a strong loss of the regular appearance of the fibrillar collagen following laser welding, which is frequently described as a full homogenization of the tissue (also called hyalinosis) [6, 9, 25]. In this context, the appearance of fibrils fused together with a drastically altered morphology is the consequence of a complete denaturation of the collagen matrix occurring at high temperatures (commonly $\geq 70\,°C$). Another frequently accompanying effect is the disruption of the cell membranes, causing leakage of the cellular material in the extracellular space. In these cases, the wound sealing mechanism has been attributed to the photocoagulation of collagen and of other intracellular proteins, which act like micro-solders or endogenous glue on laser activation, thus forming new interactions between the tissue interfaces upon cooling [25]. This mechanism is typically working when "hard" laser welding modalities are employed, as when a Ho:YAG laser ($2.01\,\mu m$ emission line) operated in continuous wave (CW) mode was tested to fuse porcine arteries at power densities up to $36.0\,W\,cm^{-2}$ [9] or the dura mater and the nerve epineurium were welded using a CO_2 laser [25].

A similar mechanism can be invoked also for laser welding by addition of protein solders at the interfaces. The temperature increase in the tissue induced by laser irradiation triggers the coagulation of the solder. Upon cooling, noncovalent interactions between the solder and the collagen matrix within the tissue are supposed to be responsible for the strength of the weld [10]. Evidence of albumin intertwining within the collagen matrix was found during microscopy analyses of specimens irradiated at temperatures above $70\,°C$ [9, 10]. Such a threshold value is in full agreement with the denaturation temperature of albumin, as reported in several spectroscopic and calorimetric studies.

9.2.2.2 Moderate Laser Welding

In other studies, less severe ("moderate") modifications of the collagen matrix were observed. In these cases, the collagen fibrils keep a recognizable structure, but appear partially swollen. Frequently, the increase in fibril caliber is accompanied by a loss of periodicity and by a splitting into fine fibrillar substructures. These effects have been reported in different cases such as in the anastomosis of rats carotid arteries by a pulsed Nd:YAG laser operated at $1.319\,\mu m$ [26] or with a diode laser at $830\,nm$ [30]. Comparable results have also been found after non-welding applications such as in the heating of tendon specimens at $60\,°C$ for $20\,min$ in a water bath [31], and in the laser-stabilization of joint capsular tissue with a holmium:YAG device [32]. In all these cases, temperatures exceeding $60\,°C$ or long exposure times promoted partial collagen denaturation characterized by an unwinding of the triple helix and a subsequent loss of fibrillar integrity. This triple-helix unraveling, followed by "interdigitation" on cooling between fibers, was indicated by Schober *et al.* as the structural basis for laser-induced welding [26].

9.2.2.3 Soft Laser Welding

A secondary role of fibrillar type I collagen in the laser welding mechanism has been pointed out in some recent studies, suggesting the involvement of some other extracellular matrix components. In these studies a combination of a low power NIR laser operated in CW mode and the topical application of NIR-absorbing dyes has been proposed for the welding of corneal tissue [8, 19, 29]. Microscopy data revealed groups of interwoven fibrils joining the sides of the cut at the weld site. In particular, transmission electron microscopy and atomic force microscopy analyses clearly indicated that type I collagen was not denatured at the operative laser energy densities employed [29]. The preservation of substantially intact, undenatured collagen fibrils in laser-welded corneal wounds is consistent with previous thermodynamic studies, which revealed temperatures below the denaturation threshold of fibrous collagen under laser irradiation. Most likely, the use of a chromophore was responsible for a controllable temperature rise confined to the area where it had been previously applied, in turn resulting in a "soft" thermal effect. In the case of low-temperature laser welding, a different model of the sealing process must be invoked. The GAGs bridges connecting collagen fibrils in the native tissue are probably broken at the characteristic temperatures of diode laser welding (50–65 °C range). The individual GAG strands, freed upon heating, subsequently create new bonds with other free strands during the cooling phase. In practice, the interwoven fibrils observed at the weld site are supposed to be connected by several newly-formed GAG bridges. The possibility of a "soft" laser welding mechanism not involving the fibrillar collagen also finds agreement with earlier studies on welded tissue extracts analyzed by gel electrophoresis [33, 34].

9.3
Temperature Control in Laser Welding Procedures

As it has been discussed above, the temperature dynamics in tissue under direct irradiation and in its surroundings strongly influence laser-induced closure of a wound. In fact, the temperature enhancement has to be high enough to induce a particular thermal process in the tissue to be sealed (Section 9.2.2), while irreversible thermal damage is to be avoided. The characteristics of the induced thermal damage depend on two main parameters, that is, the temperature and the time of exposure of the tissue at that particular temperature [23, 35]. Spatial and temporal control of the induced temperature profile would be very useful both in the optimization of the welding treatment and in its routine application.

9.3.1
Control Systems of Temperature Dynamics

The goal of temperature control is to monitor the whole thermal treatment, during and soon after laser irradiation. The control system has to be a real time, non-contact, non-invasive device, so as to enable a direct and immediate evaluation of the induced temperature without inducing any kind of damage to the tissue and without altering the measurement itself. Moreover, it has to be designed in a compact configuration to enable easy use in a surgery room. Commercially available systems that enable a real time, non-destructive detection of slow thermal processes are based on infrared detectors, which are able to monitor the thermal radiation emitted from a hot target. They are proposed in different configurations, which may be optimized for the particular application. These devices are used to control the temperature at the external surface of the biological target, while the temperature inside a biological tissue has to be estimated with theoretical models, or with destructive techniques such as with a thermocouple [36, 37]. However, the use of thermocouples provides only for a local measurement of temperature, is unsuitable to follow fast temperature dynamics, and, being in contact with the target, alters, although only slightly, the measured value.

Infrared detectors are radiometers that measure the intensity of the infrared radiation emitted by a hot target. The temperature (T) of the target is related to the intensity of the emitted infrared radiation according to the well-known Stefan–-Boltzmann law:

$$I = \varepsilon\sigma T^4 \qquad\qquad (9.1)$$

where

 ε is the emissivity of the target,
 σ is the Stefan–Boltzmann constant,
 T is the absolute temperature of the target.

Once the object parameters are known, T may be estimated. The emissivity of soft tissues is very close to that of water ($\varepsilon \sim 1$). However, several studies have been published in recent years evidencing different values of the emissivity of different biological tissues, such as the skin or the corneal surface [38, 39].

Based on these characteristics, Katzir and coworkers have developed a suitable temperature control for laser welding applications [40]: a radiometer is connected to a particular silver-halide fiber, thus enabling easy handling of the control device and a real-time monitoring of the thermal range induced on the external surface of the laser welding target. It is thus possible to control the temperature and, if necessary, to vary the laser emission.

The infrared radiation may also be collected by the use of an infrared thermo-camera [41, 42]. Commercially available devices enable real-time visualization of the surgical field. It is thus possible to control the induced temperature dynamics not only in the directly irradiated target but also in its surroundings. Modern thermo-cameras are usually connected to a computer and the data collection, visualization, and analysis may be performed by use of specific software. It is thus possible to study the spatial distribution of the temperature enhancement and its temporal evolution, with a high spatial resolution (Figure 9.5).

The main problem in the characterization of the thermal welding process is the study of the temperature evolution inside the biological target. To date, there are still no non-destructive and real time devices optimized to follow temperature evolution below the surface. The problem may be tackled only with a numerical model and solved by considering the light propagation in the tissue and the conversion of the

Figure 9.5 Visualization of the temperature enhancement in a laser welded tissue, via an infrared thermocamera. The image represents a typical user friendly interface of a commercial software package (ThermaCam Researcher, Flir Systems, Inc., Boston MA, USA): on the left there is a temperature map of the whole target, while on the right the results of some real time analyses on selected target areas are reported; at the bottom there is a temperature versus time profile as detected in a user-defined region of interest (ROI). These images are real time upgraded for immediate evaluation of the thermal processes.

transported energy into heat. It is thus very important to know the opto-thermal characteristics of the biological target and the irradiation conditions, such as the laser parameters and boundary conditions, so as to retrieve accurate temperature profiles. The numerical models may also be used to optimize the laser irradiation conditions, to reduce the thermal damage to the tissue. The model analysis is usually based on the solution of the bio-heat equation:

$$\varrho C \frac{\partial T}{\partial t} - \nabla(k\nabla T) = \varrho_b C_b \omega_b (T_b - T) + Q_{met} + Q_{ext} \tag{9.2}$$

where

ϱ and ϱ_b ($kg\,m^{-3}$) are the density of the tissue and blood respectively,
C and C_b ($J\,kg^{-1}\,K^{-1}$) are the specific heat of tissue and blood, respectively,
k ($W\,m^{-1}\,K^{-1}$) is the thermal conductivity of tissue,
ω_b (s^{-1}) is the blood perfusion rate,
T_b (K) is the arterial blood temperature,
Q_{met} ($W\,m^{-3}$) is the metabolic heat source,
Q_{ext} ($W\,m^{-3}$) is the external heat source.

In Eq. (9.2) the temperature development is described as the sum of three contributions: the blood perfusion, which is the natural way for heat dissipation through the body, the metabolic heat source describing the body's natural activity, and the contribution of the external light source. This contribution has to be developed by taking into account the optical properties of the tissue, that is, scattering and absorption, and the laser parameters, for example, laser emission modality, beam dimensions, and power [43, 44]. Several commercial software packages now offer the possibility of easily solving very complex problems of light–tissue interactions and thus of deeply understanding the basis of tissue welding [42].

9.4
Surgical Applications of Thermal Laser Welding

Despite intense research activity and preliminary tests in animal models, both *ex vivo* and *in vivo*, only a few of the proposed setups have reached the preclinical and clinical phases. In this chapter we will not review all the experimental techniques reported in the literature, which are many and have been proposed for a large variety of surgical applications (e.g., References [2, 3]). Instead, we focus mainly on a few exemplary applications, which have reached the preclinical and clinical phases. They are based on the use of a near-infrared diode laser emitting at 810 nm and the topical application of the chromophore ICG, which shows high optical absorption at the laser wavelength emission [45]. The procedure consists in a preliminary staining phase with the chromophore, followed by an irradiation phase. ICG has been chosen because of its biocompatibility, which has already favored its exploitation in several biomedical applications [46–48]. In practice, the chromophore is prepared in the form of an aqueous saturated solution of commercially available indocyanine green for biomedical applications (e.g., IC-GREEN Akorn, Buffalo Grove, IL or ICG-Pulsion Medical Systems AG, Germany). This solution is accurately positioned in the tissue area to be welded, using particular care to avoid the staining of surrounding tissues, and thus their accidental absorption of laser light. Then the wound edges are approximated and laser welding is performed under a surgical microscope.

The laser used in preclinical tests and in the clinical applications is typically an AlGaAs diode laser (e.g., Mod. WELD 800 by El.En. SpA, Italy) emitting at 810 nm and equipped with a fiber-optic delivery system.

The laser welding procedure may be optimized for a particular application by varying the irradiation conditions, such as laser power, emission modality, treatment time duration, and mutual position of the fiber tip and the tissue. In addition, different formulations of the chromophore have been proposed, including solid and semisolid ICG-containing matrices [49].

Along with the possibility of exploiting different laser emission modes, two kinds of laser welding techniques have been developed. In chronological order, the first optimized technique was continuous wave laser welding (CWLW). Non-contact, CW diode laser irradiation is used for the welding of corneal wounds, in substitution or in conjunction with traditional suturing procedures. In neurosurgery the laser welding technique offers the possibility of performing bypasses of small vessels with reduced intervention time and a less traumatic technique. Surgical procedures, thermal modeling, as well as microscopic analyses in the case of CW diode laser welding have been widely reported [8, 19, 50, 51]. The CWLW is based on the "soft" and "moderate" laser welding effect, as described above.

Later, a pulsed laser welding (PLW) procedure was set up as well, typically performed with pulse durations of 10–100 ms and energies of 10–100 mJ. Pulsed diode laser irradiation is suitable when apposition of tissue flaps is required, such as in ophthalmic surgery to secure the endothelial graft during the transplant of the corneal endothelium and in the joining of lens capsule flaps [52, 53]. In this case, "hard" laser welding induces an immediate fusion of the tissues in the laser spot area as discussed above.

9.4.1
Laser Welding in Ophthalmology

The first clinical application of laser closure of ocular tissues was in the treatment of corneal surgical wounds, in which the requirement of an immediate, watertight seal appears very useful in performing advanced ophthalmic surgeries, as well as in the treatment of accidental traumas and perforations. The development of a minimally invasive laser welding technique of transparent ocular tissues may potentially allow for the recovery of a good vision in a short post-operative time, by providing better stability of laser-welded wounds, with minimal inflammation and greater protection from infections. This could ultimately result in a better quality of life of patients and reduce hospitalization costs.

9.4.1.1 Clinical Applications in the Transplant of the Cornea

Clinical activity on the use of CWLW and PLW has been carried out by Menabuoni and coworkers at the Ophthalmic Department of the public hospital of Prato, Italy (upon approval of the Ethical Committees of the Local Health Board). The main clinical applications of CWLW are in cataract surgery [50], in penetrating keratoplasty (i.e., full thickness transplant of the cornea), and in lamellar keratoplasty (partial thickness transplant of the cornea). An improvement in lamellar (Figure 9.6) and penetrating keratoplasty was obtained by integration of the diode laser welding procedure in combination with the femtosecond laser sculpturing of corneal flaps [54], thus effectively optimizing the process for a precise and minimally invasive surgical transplant.

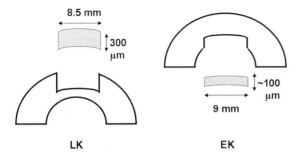

Figure 9.6 Sketch of new types of corneal transplant not involving the full thickness of the cornea. LK = lamellar keratoplasty; EK = transplant of the endothelial flap. Sizes of the donor flap in both cases are indicated.

PLW has been developed and tested for use in combination with femtosecond lasers to perform endothelial keratoplasty (i.e., transplant of the inner layer of the cornea including the endothelium layer) [55]. PLW has also been tested *ex vivo* for the closure of capsular breaks.

The CWLW procedure developed to weld human corneal tissues in penetrating keratoplasty is as follows. The donor and recipient cornea are trephined either mechanically or by use of a femtosecond laser, as suggested by the experience of the surgeon and the patients' needs. The donor cornea is then applied onto the recipient eye and secured by 8–16 interrupted stitches. Then laser welding is performed all around the perimeter of the corneal graft in substitution of the conventional continuous suturing. To this end, the chromophore solution (ICG in sterile water, 10% w/w) is placed inside the corneal cut, using an anterior chamber cannula, in an attempt to stain the walls of the cut in depth. A bubble of air is injected into the anterior chamber prior to the application of the staining solution, so as to avoid perfusion of the dye. A few minutes after the application, the solution is washed out with abundant water. The stained walls of the cut appear greenish, indicating that ICG has been absorbed by the stromal matrix. Lastly, the whole length of the cut is subjected to laser treatment. Laser energy is transferred to the tissue in a non-contact configuration, through a 300-μm core diameter fiber. A typical value of the laser power density used clinically is around 13 W cm^{-2}, which results in a good welding effect. During irradiation, the fiber tip is kept at a working distance of about 1 mm, and at an angle of 20–30° with respect to the corneal surface (side irradiation technique). This particular fiber position provides in-depth homogenous irradiation of the wound and prevents accidental irradiation of deeper ocular structures. The fiber tip is continuously moved over the tissue to be welded, with an overall laser irradiation time of about 120 s for a 25-mm cut length (the typical perimeter of a transplanted corneal button). To date, this procedure has been performed on 60 patients with very satisfactory results (Figure 9.7). The position of the apposed margins has been found to be stable over time, thus assuring optimal results in terms

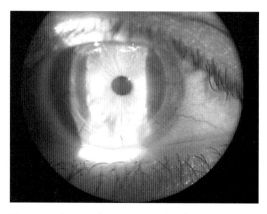

Figure 9.7 Aspect of a cornea 3 months after transplant. Diode laser welding was used to suture the perimeter of the corneal graft.

of postoperatively induced astigmatism after cataract and keratoplasty surgery. The lower number of stitches reduces the incidence of foreign body reactions, thus improving the healing process. Objective observations on treated patients have proved that the laser-welded tissues regain a good morphology (without scar formation) and pristine functionality (clarity and good mechanical load resistance).

Another important and innovative application of CWLW is in lamellar keratoplasty. A femtosecond laser is used for sculpturing a portion of the cornea. An extremely precise cut (the mechanical precision results in less than 1 µm) may be equally performed in the donor and recipient eye, or the surgeon may design a particular cut shape so as to match each patient's needs. The lamellar donor flap is then positioned in place to be sutured with the patient cornea by CWLW, with or sometimes without application of conventional suture material [56]. To this aim, the wound walls are first stained with a solution of ICG. Then the diode laser energy is transferred to the tissue in a non-contact configuration, as previously described. Intraoperatory observations and follow-up results up to 12 months indicated the formation of a smooth stromal interface, total absence of edema and inflammation, and reduced post-operative astigmatism, as compared with conventional suturing procedures. This described lamellar keratoplasty is very useful in the treatment of leukoma and central keratoconus.

PLW is used in a new transplant technique, where only the inner portion of the cornea is replaced, that is, the endothelial layer. In this case the laser welding procedure is proposed to weld the periphery of a donor endothelial flap onto the inner surface of a recipient cornea, as required in the treatment of endothelial pathologies. The potential advantage of this novel laser welding procedure is the reduction of the risk of postoperative detachment and dislocation of the transplanted endothelium from the donor cornea, which is typically high (around 30%) in conventional clinical procedures. On the other hand, since conventional stitching of the endothelial flap cannot be performed (it would require operation inside the anterior chamber), laser welding is at present the only viable alternative to secure the endothelial graft to the recipient cornea. So far, 30 patients have been subjected to PLW-assisted endothelial keratoplasty.

The optimized procedure for the welding of the endothelium is as follows. A femtosecond laser is used to prepare a 100-µm thick, 9.0-mm diameter donor corneal endothelium flap (Figure 9.8). The flap stromal (upper) side is then stained with the

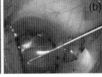

Figure 9.8 Phases of PLW-assisted endothelial transplant: (a) the endothelial flap, previously cut with the femtosecond laser, is stained with ICG only on the periphery of its stromal side; (b) the flap is then folded and introduced in the anterior chamber with forceps; (c) finally, the flap is subjected to PLW to secure it to the recipient cornea by irradiating from the outside corneal surface with the fiber probe.

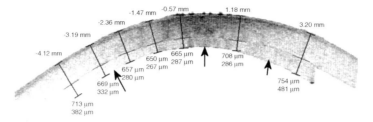

Figure 9.9 OCT image of a cornea, 1 day after the endothelium (arrows) transplant. A good approximation of the donor flap with the recipient cornea is achieved by the use of the femtosecond laser to cut the flap, while the laser welding procedure provides immediate placement of the endothelium in its position.

solution of ICG only on its periphery, taking care to avoid the accidental contamination of the lower endothelial surface. After a circular 9.0-mm diameter mechanical stripping of the recipient Descemet's membrane and endothelium, the donor flap is inserted into the anterior chamber through a 5-mm long sclerocorneal cut and positioned in contact with the inner corneal surface by an air bubble. The flap is then stabilized by means of 5–10 laser welding spots of 300-μm diameter, performed peripherally with a fiber-optic contact probe gently applied on the outer surface of the cornea. Diode laser pulses of 50 mJ and 100 ms provide effective spot welds between the donor endothelial flap and the recipient stromal surfaces. Post-operative examinations reveal complete re-absorption of the ICG within 3–5 days and excellent donor flap adhesion (Figure 9.9).

9.4.1.2 Preclinical Applications in the Closure of the Lens Capsule

One application of PLW is related to the closure of the anterior lens capsule. A successful procedure would be very useful to develop a surgical solution of refractive problems with the substitution of the lens. In fact the lens substitution could be performed with the Phaco-Ersatz intervention, by use of an anterior continuous circular capsulorhexis (with a diameter of about 1.5 mm) [57, 58]. Nuclear and cortical material is removed from the lens through the capsulorhexis and then it is substituted with a proper polymer. This may be realized by introduction of a valve to avoid possible polymer leakage. A way to fabricate this valve is to use biocompatible or donor materials, such as an anterior capsular patch, eventually extracted from a donor eye. The patch is then closed onto the recipient capsule tissue by a laser welding procedure. In fact capsular tissue is practically impossible to suture due to its elasticity and thinness (about 5–10 μm thick) (Figure 9.10).

The proposed procedure has been characterized *ex vivo* in porcine eyes, to evidence its feasibility in Phaco Ersatz interventions. A PLW procedure was performed on patches of anterior capsular tissue. The inner side of each sample is stained with a solution of ICG in sterile water 7% (w/w). Five minutes after the application, the solution is washed out with abundant water, to remove residual ICG solution. The patch is then apposed onto the external side of the anterior capsule lens of a donor eye.

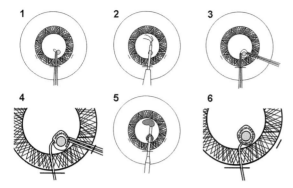

Figure 9.10 Sequence of the "lens refilling" procedure: (1) opening of a small capsulorhexis; (2) aspiration of the lens; (3) application of the stained capsular patch; (4) partial welding by PLW of the patch to create a flap-valve; (5) lens refilling with a biocompatible polymer; (6) final closure of the capsulorhexis.

In doing so, special care is taken to maintain the original orientation of the capsular sample, to match the curvature of the recipient capsular surfaces and facilitate the adhesion at the interface. The stained patch is then irradiated by means of contiguous laser spots emitted by a 200-μm core fiber, whose tip is gently pressed onto the patch surface (contact welding technique) to produce effective tissue welding. The welding procedure is performed underwater, to reproduce the anterior chamber ambient. Laser parameters able to induce optimal welding effects are pulses with energies of 30–50 mJ and durations of about 100 ms. Experimental tests proved a closure effect immediately after irradiation, with no leakage of material from the inner side of the lens bag and good mechanical resistance, comparable to that of healthy tissues (Figure 9.11). The PLW of the capsular tissue may serve as the basis for further studies for the realization of an anterior capsule valve, which may be useful for the solution of all those problems concerning the repair and the closure of the lens bag.

9.4.2
Laser Welding in Vascular Surgery

Diode laser welding is a promising tool in microvascular surgery. It has been tested *ex vivo* and *in vivo* in animal models, up to a preclinical validation of this technique. The laser welding procedure may be very useful in microsurgery and in particular in microvascular surgery, where standard suturing techniques may be problematical due to the small vessels dimensions, and where traditional methods may give rise to severe inflammatory reactions due to the presence of foreign material, ultimately leading to thrombosis and occlusion at the anastomotic site [59]. Microvascular anastomoses are commonly used in neurosurgery in the treatment of cerebral ischemia, vascular malformations, or cranial base tumors [60, 61] and are the basis of recent minimally invasive neurosurgery studies. To improve these techniques and to minimize vascular wall damage, thus improving the long-term patency, various

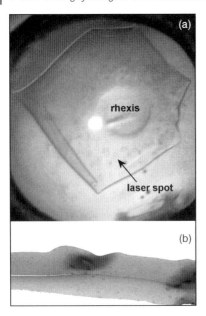

Figure 9.11 (a) Appearance of a capsulorhexis in a pig eye, which was closed by applying an ICG-stained capsular patch subjected to PLW; laser spots are clearly evident at the periphery of the patch; (b) histological slice of an anterior capsular patch (20 μm thick) laser-welded onto the anterior lens capsule of a pig eye. Effective adhesion between the two samples was accomplished with minimal heat damage (Methylene Blue staining).

alternative non-suture methods have been investigated [62–64]. The most promising approaches to laser welding of small vessels rely on the use of a diode laser emitting in the near-infrared, at around 800 nm, and a chromophore solution of ICG and other components such as hyaluronic acid (HA) and bovine serum albumin (BSA) [51, 65]. The introduction of HA in the chromophore composition has been found to increase both the viscoelasticity of the solder and the tensile strength of the resulting repairs.

Two different approaches were optimized and tested *ex vivo* and *in vivo* in animal models, evidencing the possibility of laser welding microvessels without collateral damage, while shortening the intervention times and improving healing after surgery.

The first approach described hereafter is based on the procedure developed by Puca *et al.* [41, 51] to apply a venous bypass on a carotid artery, and rests on the application of a limited number of interrupted stitches (typically 3 instead of the 10–12 used in the conventional suturing procedure) at the site of the two end-to-side anastomoses, external staining of the vessels walls with a solution of ICG and HA, and then irradiation of the wound from the outside. The other approach, developed by Frenz and coworkers, is to perform end-to-end anastomoses without any suture, [65, 66] based on the apposition of the vessel edges, external application of a solder (further improved with the implementation of a polymer scaffold), and then irradiation from

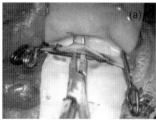 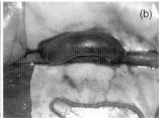

Figure 9.12 Laser welding procedure to perform a venous bypass in a carotid artery, during *in vivo* experimental surgery in a rabbit model. (a) The venous bypass is applied on the clamped carotid artery by means of three stitches on each side; the edges are stained with the ICG solution and then diode laser welding is performed. (b) When the clamps are released, the laser welded anastomoses enable high blood flow through the venous graft without any bleeding.

the inside of the vessel by means of an intraluminal balloon catheter including a side-emitting fiber.

The results of the optimization of both techniques indicated that these procedures may become safe alternatives, well suited for microvascular surgery, and find early applications in humans.

In the first approach, optimized to perform end-to-side anastomoses in carotid bypass surgery, the photosensitizer is a semisolid formulation of HA with ICG concentrations of about 7% by weight. Distal and proximal clamps are applied and end to side anastomosis is performed. The ICG solution is applied to the site of both the anastomoses by means of a plastic cannula (Figure 9.12a). After 1 min, the sites are flushed with abundant sterile water to remove the excess ICG solution. Diode laser light at 810 nm is delivered to the vessel edges by means of a 400-μm core diameter fiber-optic. The fiber is held 2–3 mm from the target and moved slowly over the vein-to-artery edges. Typical irradiation parameters are continuous diode laser power densities of 70–80 W cm^{-2} at the end of the fiber and average exposure times of 120 s to complete laser welding of the two end-to-side anastomoses, supported by three conventional stitches. The distal and proximal clamps are then removed. In the latest version of this procedure [41], the overall clamping (occlusion) time of the carotid artery was reduced to about 12 min, which is more than one-third shorter than the one necessary to perform conventional suturing or stapling procedures. This means that the laser-assisted bypass procedure may have a potential to significantly reduce the risk of cerebral complications due to long lasting occlusions of the carotid artery.

The efficacy of the procedure is immediately verified: when the clamps are released, laser-assisted anastomoses typically show no bleeding, whereas conventional anastomoses present some bleeding. The patency of the bypass may be verified *in vivo*, both by Doppler flow analysis and by direct observation of the blood flow (Figure 9.12b). Histological, immunohistochemical, and SEM analyses were performed in a follow-up evaluation study on animal models, to study the welded tissue and to compare the effect of laser welding with conventional suturing. It was demonstrated that the laser-assisted suturing technique provides better healing processes. The morphological evaluation of the vessels walls in sutured bypasses indicates the presence of marked traumatic and inflammatory phenomena, localized in the perisutural area, and, due to the traumatic suturing event, the foreign body reaction and an ischemic-necrotic damage, produced by the tension exerted by the suture in the area around the surgical wire. In the laser-assisted procedure all these events are reduced. In this case hyalinosis of the vascular wall induced by collagen fusion is a characteristic pathological fingerprint. Collagen in the media layer turns into a hyaline matrix, due to the laser-induced photothermal effect. SEM and immunohistochemical analyses confirmed this trend.

In the second approach, microvascular welding is implemented by irradiating the wound from the intraluminal side, by use of a balloon catheter as a temporary stent (Figure 9.13). Precise alignment and adaptation of the vascular stumps is thus provided, without the need of stay sutures. In addition, handling would no longer be influenced by spatial constraints around the anastomotic site, when the laser source is placed within the vascular lumen. Laser irradiation is provided by a suitably designed laser fiber, implemented into the center of a balloon catheter, around which the vascular stumps may be laid for approximation and transmural irradiation focused on an externally applied solder. This set up has been tested both *ex vivo* and *in vivo* in animal models. The laser fiber is based on a 400-μm low-OH quartz

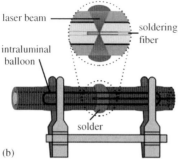

Figure 9.13   (a) Intraoperative image showing the two vessel stumps (cut by an arrow) fixed on the dilated balloon catheter before applying the solder. The right image edge shows a millimeter scale. (b) Schematic of the laser-assisted vessel soldering equipment. Courtesy of Martin Frenz.

fiber and a cone-shaped silver mirror at its tip, producing a 360° irradiation ring of homogeneous irradiance over a length of about 2 mm perpendicular to the fiber axis. A continuous wave 808-nm diode laser (DL50, Fisba Optik, St. Gallen, Switzerland) was coupled into the soldering fiber. Irradiances were in the range 1.5–5 W cm^{-2}. To enhance laser light absorption a solder was used to stain the external anastomotic site. The solder consisted of 25% BSA, 0.5% HA, 0.1% ICG, and sterilized water. In successive studies the solder was optimized with a biocompatible polymer scaffold, to prevent excessive loss of the solder and to increase the absorbed energy density at the point of interest. The scaffold was a dry polycaprolactone (PCL), doped with BSA and ICG, and named BIP-scaffold [65]. This PCL polymer was chosen because of its tested compatibility (it has received FDA approval) and its porous structure, which enables maintenance of the liquid solder in place during the thermal process. In addition, PCL has a low melting point, at around 60 °C. A more homogenous and therefore stronger polymer–BSA–tissue-interaction was thus obtained by concomitant melting of the polymer at the interface. Moreover, different thicknesses and shapes of this soldering scaffold can be realized for an optimal adaptation to the surgical requirements. *Ex vivo* tests were performed recently to optimize irradiation parameters so as to obtain maximum tensile strength and minimum tissue damage, thus improving the intraluminal laser soldering technique. The optimized irradiance was ca. 7 W cm^{-2}, with a time duration of 60 s. Histological analyses clearly showed that the melted PCL creates a smooth seal at the interface between scaffold and tissue. This intraluminal technique with BIP-scaffolds thus promises to become a useful tool in microvascular surgery.

9.5
Future Perspectives

Current efforts to extend the range of clinical applications of laser welding include in particular the development of novel concepts of exogenous chromophores.

Conventional exogenous chromophores of common use in surgical practices are organic dyes, such as the ICG. Despite the experimental and clinical achievements discussed in the paragraphs above, these molecules are less than ideal and exhibit poor performances in several critical respects [67, 68]. For instance, the light extinction efficiency and photochemical stability of solutions of organic dyes are limited. Their conformation and optical response depend strongly on their biochemical environment and temperature, and typically deteriorate rapidly over time [69]. Their biochemical behavior is hardly versatile, due to their poor possibility of conjugation with additional functional moieties. Among the other drawbacks, these limitations hinder, for example, a better control over their perfusion and delivery to the biological targets of interest, such as, for example, topically through biocompatible insets and systemically through intravenous injection. Therefore, there is growing interest in the introduction of alternative concepts of exogenous chromophores, which may improve the feasibility of laser welding and other kinds of diagnostic and therapeutic applications.

Recent progress in the design and synthesis of novel colloidal suspensions has produced various nanoparticles with excellent optical response, which hold the potential to outclass organic dyes presently in use. Examples include hybrid materials composed of ICG molecules embedded within organic or inorganic nanoparticles [69]. These systems rest on ordinary concepts of photothermal conversion, realize high local concentrations of the organic dyes, exhibit better stability in the biological environment, and may become conjugated with further functional molecules.

In addition to these hybrid materials, there exist alternative proposals of exogenous chromophores, which rely on radically innovative concepts of photothermal conversion. Here we mention metal oxide and metal nanoparticles with magnetic and plasmonic resonances, respectively. To this broad class of systems belongs for instance various non-spherical gold nanoparticles, whose light extinction originates from the excitation of collective charge oscillations at frequencies in the NIR window of principal interest [70, 71]. These colloidal suspensions have several advantages over the organic dyes in use [67, 72]. Their light extinction spectra may be tuned throughout the NIR window with accessible parameters, such as their size and shape distributions. Remarkable features include exceptional molar extinction coefficients (higher by up to five orders of magnitude than that of ICG in the case of so-called gold nanorods) [67], excellent stability in the body even at high irradiation levels and temperatures, and also the possibility of flexible conjugation with additional bio-chemical functionalities [73]. Various non-spherical gold nanoparticles with attractive optical response in the NIR window have been synthesized over recent years, including pure metal nanoparticles such as, for example, so-called gold nanorods [70] and gold nanocages [74], as well as core–metal shell nanoparticles with, for example, a silica core [75] or a carbon nanotube core [76].

Numerous biomedical applications of such nanoparticles have been proposed in the scientific literature, with particular focus on unconventional approaches for the diagnosis and therapeutic treatment of cancer (e.g., References [75, 77]). In contrast much less effort has been devoted to the closure of wounds.

In this context, for instance, Gobin *et al.* have demonstrated the use of spherical plasmonic nanoparticles composed of a silica core and a gold shell (100 nm overall diameter) to sensitize an albumin solder and successfully seal muscular (*ex vivo*) and cutaneous (*in vivo*, in rat) lesions upon laser irradiation at 800–820 nm [78]. Analogous results were achieved by Bregy *et al.*, who employed an albumin solder enriched with spherical superparamagnetic iron oxide nanoparticles (15 nm diameter) to fuse blood vessels (*ex vivo*) on application of 170 m radio waves, which penetrate very deep through the whole body [79].

More recently, our research group has proved the use of biocompatible cylindrical gold nanoparticles, named gold nanorods, with plasmon resonances at ∼800 nm to assist the laser welding of patches of porcine eye lens capsules [80]. This approach differs from previous examples in the direct application of the nanoparticles stain onto the native tissues, and so in the direct stimulation of a photothermal modification within the native tissues (as opposed to a foreign solder). In this particular context, replacement of the organic dyes with nanoparticles (∼10–20 nm diameter, ∼40–80 nm long) may be further beneficial to gain a better localization of the

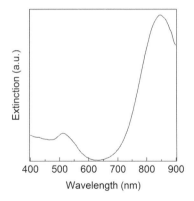

Figure 9.14 Extinction spectrum of a colloidal suspension of gold nanorods.

exogenous chromophores, due to the relatively large size of nanoparticles and, consequently, slower diffusivity through the collagen fibrils network.

In the example reported by our research group, laser welding tests simulated transplants of patches of the eye lens capsule from a donor onto the eye lens capsule of a recipient, which may become of interest in the realm of ophthalmic surgery, for example, in problematical operations such as the Phaco-Ersatz eye lens refilling. The interface between the donor and the recipient tissues was stained with gold nanorods, such as those displayed in Figures 9.14 and 9.15.

After application of the gold nanorods stain, 810 nm diode laser light was delivered through a 300-μm diameter optical fiber. The power was given spot-wise in single pulses in the 10–100 ms time regime. Under these conditions, reproducible welds were realized, with a mechanical strength similar to the native tissues, below ~100 J cm^{-2} light fluence, which is several times too low to induce any detectible effects in eye lens capsules with no stain. Microscopic evaluation of the weld sites proved an excellent localization of the photothermal damage. Although collagen denaturation and shrinkage was observed at the interface within the spots under direct laser irradiation (Figure 9.16), the normal appearance and optical functionality of the native tissue was recovered in a few tens of microns. The results achieved by our research group prove the feasibility and potential of these novel technologies in direct laser welding of connective tissues.

As one further perspective, the wealth of concepts to modify the surface of metal nanoparticles may enable, for example, their inclusion into durable and

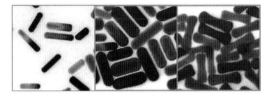

Figure 9.15 Transmission electron micrographs (200 × 200 nm^2) of gold nanorods of different average size and shape.

Figure 9.16 Histological section of a transplant of porcine eye lens capsules realized under the conditions described in the main text. The hematoxylin/eosin stain (dark gray) emphasizes the collagen denaturation at the weld site.

biocompatible films and gels, which may be loaded with additional drugs and factors to optimize the delivery of the exogenous chromophores and the repair after the surgery.

Currently, the replacement of organic dyes with laser activated nanoparticles is an inspiring possibility that is still quite far from the clinical arena. As a very recent proposal, there still exists limited experimental proof of its actual potential in the laser welding of connective tissues. Additional aspects of practical concern include the biocompatibility and sustainability of laser-activated nanoparticles [81]. In short, the new and thriving context of nanomedicine is quickly penetrating laser welding technology as well, and is certainly among its most pioneering frontiers. We foresee exciting evolution in the near future, possibly along some of the guidelines sketched in this section.

Acknowledgment

The authors wish to thank Luca Menabuoni and Ivo Lenzetti from the Ophthalmic Department of the Hospital of Prato, Italy, who set up the clinical applications of corneal laser welding; Alfredo Puca, Alessio Albanese, and Giuseppe Esposito from the Institute of Neurosurgery of the Catholic University in Rome, Italy, who performed the laser-assisted bypass surgery; and Martin Frenz from the Institute of Applied Physics of University of Bern, Switzerland, for fruitful discussions and for permission to use Figure 9.13.

References

1 Jain, K.K. and Gorisch, W. (1979) Repair of small blood vessels with the neodymium-YAG laser: a preliminary report. *Surgery*, **85** (6), 684–688.

2 McNally, K.M. (2003) Laser tissue welding, in *Biomedical Photonics Handbook* (ed. T. Vo-Dihn), CRC Press, Boca Raton, Ch. 39, pp. 1–45.

3 Pini, R., Rossi, F., Matteini, P., and Ratto, F. (2008) Laser tissue welding in minimally invasive surgery and

microsurgery, in *Biophotonics: Biological and Medical Physics, Biomedical Engineering* (eds L. Pavesiand P.M. Fauchet), Springer, pp. 275–299.

4 Mulroy, L., Kim, J., Wu, I., Scharper, P., Melki, S.A., Azar, D.T., Redmond, R.W., and Kochevar, I.E. (2000) Photochemical keratodesmos for repair of lamellar corneal incisions. *Invest. Ophthalmol. Vis. Sci.*, **41** (11), 3335–3340.

5 Kopchok, G.E., White, R.A., White, G.H., Fujitani, R., Vlasak, J., Dykhovsky, L., and Grundfest, W.S. (1988) CO_2 and argon laser vascular welding: acute histologic and thermodynamic comparison. *Lasers Surg. Med.*, **8**, 584–588.

6 DeCoste, S.D., Farinelli, W., Flotte, T., and Anderson, R.R. (1992) Dye-enhanced laser welding for skin closure. *Lasers Surg. Med.*, **12** (1), 25–32.

7 Poppas, D.P. and Scherr, D.S. (1998) Laser tissue welding: a urological surgeon's perspective. *Haemophilia*, **4** (4), 456–462.

8 Rossi, F., Pini, R., Menabuoni, L., Mencucci, R., Menchini, U., Ambrosini, S., and Vannelli, G. (2005) Experimental study on the healing process following laser welding of the cornea. *J. Biomed. Opt.*, **10** (2), 024004.

9 Ott, B., Zuger, B.J., Erni, D., Banic, A., Schaffner, T., Weber, H.P., and Frenz, M. (2001) Comparative in vitro study of tissue welding using a 808 nm diode laser and a Ho:YAG laser. *Lasers Med. Sci.*, **16** (4), 260–266.

10 McNally, K.M., Sorg, B.S., Welch, A.J., Dawes, J.M., and Owen, E.R. (1999) Photothermal effects of laser tissue soldering. *Phys. Med. Biol.*, **44** (4), 983–1002.

11 Kirsch, A.J., Cooper, C.S., Gatti, J., Scherz, H.C., Canning, D.A., Zderic, S.A., and Snyder, H.M. (2001) Laser tissue soldering for hypospadias repair: results of a controlled prospective clinical trial. *J. Urol.*, **165** (2), 574–577.

12 Lauto, A., Hamawy, A.H., Phillips, A.B., Petratos, P.B., Raman, J., Felsen, D., Ko, W., and Poppas, D.P. (2001) Carotid artery anastomosis with albumin solder and near infrared lasers: a comparative study. *Lasers Surg. Med.*, **28** (1), 50–55.

13 Sorg, B.S. and Welch, A.J. (2002) Tissue welding with biodegradable polymer films-demonstration of acute strength reinforcement *in vivo*. *Lasers Surg. Med.*, **31** (5), 339–342.

14 Poppas, D.P., Massicotte, J.M., Stewart, R.B., Roberts, A.B., Atala, A., Retik, A.B., and Freeman, M.R. (1996) Human albumin solder supplemented with TGF-beta 1 accelerates healing following laser welded wound closure. *Lasers Surg. Med.*, **19** (3), 360–368.

15 Lauto, A., Stoodley, M., Marcel, H., Avolio, A., Sarris, M., McKenzie, G., Sampson, D.D., and Foster, L.J. (2007) In *vitro* and in vivo tissue repair with laser-activated chitosan adhesive. *Lasers Surg. Med.*, **39**, 19–27.

16 Scott, J.E. and Haigh, M. (1988) Identification of specific binding sites for keratan sulphate proteoglycans and chondroitin-dermatan sulphate proteoglycans on collagen fibrils in cornea by the use of cupromeronic blue in 'critical-electrolyte-concentration' techniques. *Biochem. J.*, **253**, 607–610.

17 Scott, J.E. (1992) Morphometry of cupromeronic blue-stained proteoglycan molecules in animal corneas, versus that of purified proteoglycans stained in *vitro*, implies that tertiary structures contribute to corneal ultrastructure. *J. Anat.*, **180**, 155–164.

18 Matteini, P., Dei, L., Carretti, E., Volpi, N., Goti, A., and Pini, R. (2009) Structural behavior of hyaluronan under high concentration conditions. *Biomacromolecules*, **10**, 1516–1522.

19 Matteini, P., Rossi, F., Menabuoni, L., and Pini, R. (2007) Microscopic characterization of collagen modifications induced by low-temperature diode-laser welding of corneal tissue. *Lasers Surg. Med.*, **39**, 597–604.

20 Allain, J.C., Le Lous, M., Cohen-Solal, L., Bazin, S., and Maroteaux, P. (1980) Isometric tensions developed during the hydrothermal swelling of rat skin. *Conn. Tissue Res.*, **7**, 127–133.

21 Le Lous, M., Flandin, F., Herbage, D., and Allain, J.C. (1982) Influence of collagen denaturation on the chemorheological properties of skin, assessed by differential

scanning calorimetry and hydrothermal isometric tension measurement. *Biochim. Biophys. Acta*, **717**, 295–300.

22 Kampmeier, J., Radt, B., Birngruber, R., and Brinkmann, R. (2000) Thermal and biomechanical parameters of porcine cornea. *Cornea*, **19** (3), 355–363.

23 Brinkmann, R., Radt, B., Flamm, C., Kampmeier, J., Koop, N., and Birngruber, R. (2000) Influence of temperature and time on thermally induced forces in corneal collagen and the effect on laser thermokeratoplasty. *J. Cataract Refract Surg.*, **26**, 744–754.

24 Savage, H.E., Halder, R.K., Kartazayeu, U., Rosen, R.B., Gayen, T., McCormick, S.A., Patel, N.S., Katz, A., Perry, H.D., Paul, M., and Alfano, R.R. (2004) NIR laser tissue welding of in vitro porcine cornea and sclera tissue. *Lasers Surg. Med.*, **35** (4), 293–303.

25 Menovsky, T., Beek, J.F., and van Gemert, M.J.C. (1996) Laser tissue welding of dura mater and peripheral nerves: a scanning electron microscopy study. *Lasers Surg. Med.*, **19** (2), 152–158.

26 Schober, R., Ulrich, F., Sander, T., Durselen, H., and Hessel, S. (1986) Laser-induced alteration of collagen substructure allows microsurgical tissue welding. *Science*, **232** (4756), 1421–1422.

27 Tan, H.Y., Teng, S.W., Lo, W., Lin, W.C., Lin, S.J., Jee, S.H., and Dong, C.Y. (2005) Characterizing the thermally induced structural changes to intact porcine eye, part 1: second harmonic generation imaging of cornea stroma. *J. Biomed. Opt.*, **10**, 540191–540195.

28 Matteini, P., Ratto, F., Rossi, F., Cicchi, R., Stringari, C., Kapsokalyvas, D., Pavone, F.S., and Pini, R. (2009) Photothermally-induced disordered patterns of corneal collagen revealed by SHG imaging. *Opt. Express*, **17**, 4868–4878.

29 Matteini, P., Sbrana, F., Tiribilli, B., and Pini, R. (2009) Atomic force microscopy and transmission electron microscopy analyses on low-temperature laser welding of the cornea. *Lasers Med. Sci.*, **24**, 667–671.

30 Tang, J., Godlewski, G., Rouy, S., and Delacretaz, G. (1997) Morphologic changes in collagen fibers after 830 nm diode laser welding. *Lasers Surg. Med.*, **21** (5), 438–443.

31 Anderson, R.R., Lemole, G.M., Kaplan, R., Solhpour, S., Michaud, N., and Flotte, T. (1994) Molecular mechanism of thermal tissue welding. *Lasers Surg. Med.*, **6** (Suppl.), 56.

32 Hayashi, K., Thabit, Gr., Bogdanske, J.J., Mascio, L.N., and Markel, M.D. (1996) The effect of nonablative laser energy on the ultrastructure of joint capsular collagen. *Arthroscopy*, **12**, 474–481.

33 Murray, L.W., Su, L., Kopchok, G.E., and White, R.A. (1989) Crosslinking of extracellular matrix proteins: a preliminary report on a possible mechanism of argon laser welding. *Lasers Surg. Med.*, **9** (5), 490–496.

34 Guthrie, C.R., Murray, L.W., Kopchok, G.E., Rosenbaum, D., and White, R.A. (1991) Biochemical mechanisms of laser vascular tissue fusion. *J. Invest. Surg.*, **4** (1), 3–12.

35 Niemz, M. (2007) *Laser-Tissue Interactions: Fundamentals and Applications*, Springer Verlag.

36 Soller, E.C., Hoffman, G.T., and McNally-Heintzelman, K.M. (2003) Optimal parameters for arterial repair using light-activated surgical adhesives. *Biomed. Sci. Instrum.*, **39**, 18–23.

37 Donnenfeld, E.D., Olson, R.J., Solomon, R., Finger, P.T., Biser, S.A., Perry, H.D., and Doshi, S. (2003) Efficacy and wound temperature gradient of White Star phacoemulsification through a 1.2 mm incision. *J. Cataract Refract Surg.*, **29**, 1097–1100.

38 Scott, J.A. (1988) A finite element model of heat transport in the human eye. *Phys. Med. Biol.*, **33**, 227–241.

39 Togawa, T. (1989) Non-contact skin emissivity: measurement from reflectance using step change in ambient radiation temperature. *Clin. Phys. Physiol. Meas.*, **10**, 39–48.

40 Sa'ar, A., Moser, F., Skselrod, S., and Katzir, A. (1986) Infrared optical properties of polycrystalline silver halide fibers. *Appl. Phys. Lett.*, **49**, 305–307.

41 Puca, A., Esposito, G., Albanese, A., Maira, G., Rossi, F., and Pini, R. (2009) Minimally occlusive laser vascular anastomosis

(MOLVA): experimental study. *Acta Neurochir.*, **151**, 363–368.

42 Rossi, F., Pini, R., and Menabuoni, L. (2007) Experimental and model analysis on the temperature dynamics during diode laser welding of the cornea. *J. Biomed. Opt.*, **12** (1), 014031.

43 Sagi, A., Shitzer, A., and Katzir, A. (1992) Heating of biological tissue by laser irradiation: theoretical model. *Opt. Eng.*, **31**, 1417–1424.

44 Welch, A.J. (1984) The thermal response of laser irradiated tissue. *IEEE J. Quantum Electron.*, **20**, 1471–1481.

45 Landsman, M.L.J., Kwant, G., Mook, G.A., and Zijlstra, W.G. (1976) Light-absorbing properties, stability, and spectral stabilization of indocyanine green. *J. Appl. Physiol.*, **40** (4), 575–583.

46 Yannuzzi, L.A., Ober, M.D., Slakter, J.S., Spaide, R.F., Fisher, Y.L., Flower, R.W., and Rosen, R. (2004) Ophthalmic fundus imaging: today and beyond. *Am. J. Ophthalmol.*, **137** (3), 511–524.

47 Holley, G.P., Alam, A., Kiri, A., and Edelhauser, H.F. (2002) Effect of indocyanine green intraocular stain on human and rabbit corneal endothelial structure and viability. An *in vitro* study. *J. Cataract Refract Surg.*, **28** (6), 1027–1033.

48 Kamolza, L.-P., Andel, H., Haslik, W., Donner, A., Winter, W., Meissl, G., and Frey, M. (2003) Indocyanine green video angiographies help to identify burns requiring operation. *Burns*, **29**, 785–791.

49 Chetoni, P., Burgalassi, S., Monti, D., Tampucci, S., Rossi, F., Pini, R., and Menabuoni, L. (2007) Healing of rabbits' cornea following laser welding: effect of solid and semisolid formulations containing indocyanine green. *J. Drug Del. Sci. Tech.*, **17**, 25–31.

50 Menabuoni, L., Pini, R., Rossi, F., Lenzetti, I., Yoo, S.H., and Parel, J.-M. (2007) Laser-assisted corneal welding in cataract surgery: a retrospective study. *J. Cataract Refract. Surg.*, **33**, 1608–1612.

51 Puca, A., Albanese, A., Esposito, G., Maira, G., Tirpakova, B., Rossi, G., Mannocci, A., and Pini, R. (2006) Diode laser-assisted carotid bypass surgery: an experimental study with morphological and immunohistochemical evaluations. *Neurosurgery*, **59** (6), 1286–1294.

52 Rossi, F., Matteini, P., Ratto, F., Menabuoni, L., Lenzetti, I., and Pini, R. (2008) Laser tissue welding in ophthalmic surgery. *J. Biophoton.*, **1**, 331–342.

53 Pini, R., Rossi, F., Menabuoni, L., Lenzetti, I., Yoo, S., and Parel, J.M. (2008) A new technique for the closure of the lens capsule by laser welding. *Ophthalmic Surg. Lasers Imaging.*, **39** (3), 260–261.

54 Menabuoni, L., Lenzetti, I., Cortesini, L., Pini, R., and Rossi, F. (2009) Technical improvements in PK performed with the combined use of femtosecond and diode lasers. *Invest. Ophthalmol. Vis. Sci.*, **50**, E-Abstract 2220.

55 Menabuoni, L., Lenzetti, I., Rutili, T., Rossi, F., Pini, R., Yoo, S.H., and Parel, J.M. (2007) Combining femtosecond and diode lasers to improve endothelial keratoplasty outcome. A preliminary study. *Invest. Ophthalmol. Vis. Sci.*, **45**, E-Abstract 4711.

56 Menabuoni, L., Pini, R., Fantozzi, M., Susini, M., Lenzetti, I., and Yoo, S.H. (2006) "All-laser" sutureless lamellar keratoplasty (ALSL-LK): a first case report. *Invest. Ophthalmol. Vis. Sci.*, **47**, E-Abstract 2356.

57 Haefliger, E. and Parel, J.-M. (1994) Accommodation of an endocapsular silicone lens (Phaco-Ersatz) in the aging rhesus monkey. *J. Refract Corneal Surg.*, **10** (5), 550–555.

58 Parel, J.M., Gelender, H., Trefers, W.F., and Norton, E.W. (1986) Phaco-Ersatz: cataract surgery designed to preserve accommodation. *Graefes Arch. Clin. Exp. Ophthalmol.*, **224**, 165–173.

59 Zeebregts, C.J., Heijmen, R.H., van den Dungen, J.J., and van Schilfgaarde, R. (2003) Non-suture methods of vascular anastomosis. *Br. J. Surg.*, **90** (3), 261–271.

60 Lawton, M.T., Hamilton, M.G., Morcos, J.J., and Spetzler, R.F. (1996) Revascularization and aneurysm surgery: current techniques, indications, and outcome. *Neurosurgery*, **38** (1), 83–92.

61 Sekhar, L.N. and Kalavakonda, C. (2002) Cerebral revascularization for aneurysms and tumors. *Neurosurgery*, **50** (2), 321–331.

62 Aksik, I.A., Kikut, R.P., and Apshkalne, D.L. (1986) Extraintracranial anastomosis performed by means of biological gluing materials: experimental and clinical study. *Microsurgery*, **7** (1), 2–8.

63 Ang, E.S., Tan, K.C., Tan, L.H., Ng, R.T., and Song, I.C. (2001) 2-Octylcyanoacrylate-assisted microvascular anastomosis: comparison with a conventional suture technique in rat femoral arteries. *J. Reconstr. Microsurg.*, **17** (3), 193–201.

64 Falconer, D.P., Lewis, T.W., Lamprecht, E.G., and Mendenhall, H.V. (1990) Evaluation of the Unilink microvascular anastomotic device in the dog. *J. Reconstr. Microsurg.*, **6** (3), 215–222.

65 Bregy, A., Bogni, S., Bernau, V.J., Vajtai, I., Vollbach, F., Petri-Fink, A., Constantinescu, M., Hofmann, H., Frenz, M., and Reinert, M. (2008) Solder doped polycaprolactone scaffold enables reproducible laser tissue soldering. *Lasers Surg. Med.*, **40**, 716–725.

66 Ott, B., Constantinescu, M., Erni, D., Banic, A., Schaffner, T., and Frenz, M. (2004) Intraluminal laser light source and external solder: in vivo evaluation of a new technique for microvascular anastomosis. *Lasers Surg. Med.*, **35**, 312–316.

67 Jain, P.K., Lee, K.S., El-Sayed, I.H., and El-Sayed, M.A. (2006) Calculated absorption and scattering properties of gold nanoparticles of different size, shape, and composition: applications in biological imaging and biomedicine. *J. Phys. Chem. B*, **110**, 7238.

68 Wang, F., Tan, W.B., Zhang, Y., Fan, X., and Wang, M. (2006) Luminescent nanomaterials for biological labelling. *Nanotechnology*, **17**, R1–R13.

69 Saxena, V., Sadoqi, M., and Shao, J. (2004) Indocyanine green-loaded biodegradable nanoparticles: preparation, physicochemical characterization and in vitro release. *Int. J. Pharm.*, **278**, 293–301.

70 Pérez-Juste, J., Pastoriza-Santos, I., Liz-Marzán, L.M., and Mulvaney, P. (2005) Gold nanorods: synthesis, characterization and applications. *Coord. Chem. Rev.*, **249**, 1870–1901.

71 Jain, P.K., Huang, X., El-Sayed, I.H., and El-Sayed, M.A. (2008) Noble metals on the nanoscale: optical and photothermal properties and some applications in imaging, sensing, biology, and medicine. *Acc. Chem. Res.*, **41**, 1578–1586.

72 Hu, M., Chen, J., Li, Z.Y., Au, L., Hartland, G.V., Li, X., Marqueze, M., and Xia, Y. (2006) Gold nanostructures: engineering their plasmonic properties for biomedical applications. *Chem. Soc. Rev.*, **35**, 1084–1094.

73 Daniel, M.C. and Astruc, D. (2004) Gold nanoparticles: assembly, supramolecular chemistry, quantum-size-related properties, and applications toward biology, catalysis, and nanotechnology. *Chem. Rev.*, **104**, 293–346.

74 Chen, J., McLellan, J.M., Siekkinen, A., Xiong, Y., Li, Z.Y., and Xia, Y. (2006) Facile synthesis of gold-silver nanocages with controllable pores on the surface. *J. Am. Chem. Soc.*, **128**, 14776–14777.

75 Hirsch, L.R., Gobin, A.M., Lowery, A.R., Tam, F., Drezek, R.A., Halas, N.J., and West, J.L. (2006) Metal nanoshells. *Ann. Biomed. Eng.*, **34**, 15–22.

76 Kim, J.-W., Galanzha, E.I., Shashkov, E.V., Moon, H.-M., and Zharov, V.P. (2009) Golden carbon nanotubes as multimodal photoacoustic and photothermal high-contrast molecular agents. *Na. Nanotech.* doi: 10.1038/nnano.2009.231.

77 Jain, P.K., El-Sayed, I.H., and El-Sayed, M.A. (2007) Au nanoparticles target cancer. *Nano Today*, **2**, 18–29.

78 Gobin, A.M., O'Neal, D.P., Watkins, D.M., Halas, N.J., Drezek, R.A., and West, J.L. (2005) Near infrared laser-tissue welding using nanoshells as an exogenous absorber. *Lasers Surg. Med.*, **37** (2), 123–129.

79 Bregy, B., Kohler, A., Steitz, B., Petri-Finkb, A., Bognid, S., Alfieri, A., Münker, M., Vajtaif, I., Frenz, M., Hofmann, H., and Reinert, M. (2008) Electromagnetic tissue fusion using superparamagnetic iron oxide nanoparticles: first experience with rabbit aorta. *Open Surgery J.*, **2**, 3–9.

80 Ratto, F., Matteini, P., Rossi, F., Menabuoni, L., Tiwari, N., Kulkarni, S.K., and Pini, R. (2009) Photothermal effects in connective tissues mediated by laser-activated gold nanorods. *Nanomedicine*, **5**, 143–151.

81 Lewinski, N., Colvin, V., and Drezek, R. (2008) Cytotoxicity of nanoparticles. *Small*, **4**, 26–49.

82 Barak, A., Eyal, O., Rosner, M., Belotserkousky, E., Solomon, A., Belkin, M., and Katzir, A. (1997) Temperature-controlled CO_2 laser tissue welding of ocular tissues. *Surv. Ophthalmol.*, **42** (Suppl 1), S77–S81.

Conclusions

Francesco S. Pavone

This book has shown how the implementation of imaging capability with the possibility of optically manipulating a sample allows for new kinds of applications in cell biology.

After an initial explanation of multiphoton imaging, Professor König and Dr. Aisada Uchugonova have shown the capabilities of imaging and nanoprocessing with femtosecond lasers (Chapter 1). When using multiphoton imaging they have demonstrated the detection of metabolic activity, opening up new perspectives for the study of cell physiology also *in vivo*. This will allow us to connect the cell pathway with more general tissue physiology, related to interactions of a group of cells. This kind of analysis has been applied to stem cells, in particular. Of course, more studies on various stem cells under different types of differentiation are required to understand the exact metabolism during the differentiation process. It will also be possible to use second-harmonic generation imaging, in the case of osteogenic and chondrogenic differentiation, to detect the biosynthesis of collagen.

Moreover, the use of very short pulse width operation has demonstrated the capability of performing some nanoprocessing on living cells.

The potential use of such low-power systems opens the way for the manufacture of ultra-compact, low-power laser systems for nanoprocessing as well as for imaging at a lower price than current femtosecond laser microscopes.

Stem cells are now at a stage from "bench to bedside" for the treatment of myocardial infarction, neurological diseases such as Alzheimer's and Parkinson, diabetes, and cancer. Current stem cell therapy is based on adult stem cells, where the recruitment of the limited number and their restricted locations is extremely difficult. Furthermore, the stem cells of a patient with genetically based diseases will also carry the genetic effect. It is hoped that multiphoton imaging and nanoprocessing tools can be employed to trace and to manipulate the stem cells as well as to monitor and to influence the differentiation process. However, future work on many different stem cells within their native tissue environment has to be conducted to find out if the rare stem cells can be identified due to their characteristic autofluorescence behavior.

Of course, the ultimate goal will be *in vivo* measurements, such as stem cell migration, homing processes, or differentiation. This is already a reality, and future

Laser Imaging and Manipulation in Cell Biology. Edited by Francesco S. Pavone
Copyright © 2010 WILEY-VCH Verlag GmbH & Co. KGaA, Weinheim
ISBN: 978-3-527-40929-7

Figure 1 *In vivo* dermoscope(MTPflex from JenLab GmbH).

applications will allow the realization of multiphoton microscopes to work *in vivo* (Figure 1).

Also in the near future, two-photon microendoscopes will be available and be employed to trace stem cells inside the body.

As for the nanoprocessing operation with a femtosecond laser, the optical generation of transient nanoholes will open up a novel way to deliver not only foreign DNA but also other molecules and chemicals such as RNA, recombinant proteins, nanoparticles, and drugs into the living cell, without destructive collateral effects.

Future engineering work may result in the development of fiber-based systems with automatic miniaturized high-throughput femtosecond laser systems.

To demonstrate the evolution of living cell nanoprocessing towards *in vivo* measurements, our group has operated this optical manipulation in living organisms.

We have shown how multiphoton absorption has reinvigorated this area of research and become a useful tool for the selective disruption and dissection of cellular structures in living cells and *in vivo*. By varying the laser repetition rate, pulse energy, number of pulses irradiated, and the focusing conditions, femtosecond nanosurgery can be performed with different interaction regimes and work both on the subcellular and the tissue level with unprecedented specificity. As illustrated by the few examples described in Chapter 2, the impact of multiphoton nanosurgery on modern biology (especially, but not only, cell biology and neurobiology) has already been remarkable. The capability of combining visualization with selective and direct disruption of intracellular structures offers an extraordinary tool for the investigation of many vital processes, including cellular division, locomotion, and cytoskeletal plasticity. The combination of multiphoton nanosurgery and *in vivo* imaging represents a promising tool for probing and disrupting neuronal circuits. The potential of using this precise optical method to perturb individual synapses cannot be overstated. Using multiphoton nanosurgery, the synaptic organization of the brain can now be teased apart *in vivo* to understand the microcircuitry of neuronal networks. Microscopy, therefore, has clearly expanded from its initial and very important role in pure imaging (yet improving impressively also in that area) to become a fully comprehensive and versatile tool in biological investigation.

Future applications of this methodology will be expanded to study the reaction of the central nervous system upon laser focal damaging when treated with pharmacological compounds. This will open up interesting perspectives for the study of neurodegenerative disease, for example, and their pharmacological treatment.

Many kinds of damages can be produced with optical disruption, such as dendrite cutting, single spine ablation, induced strokes on blood vessel, and whole cell ablation (Figure 2). In all these phenomena, different reactions of the system will be involved, and many different pathologies will be studied.

Spine plasticity, for example, is directly correlated to the capability of the system to realize electrical connections, and the motility of immune cells after damage is directly connected to the capability of the immune system to react to an external "attack," or to the state of defense of the circuit. Stem cell homing after damage is, for example, a completely open field with many applications in "repair" operations after damage, induced trauma after accident, or neurodegenerative pathologies.

The physical mechanism of ablation plays an important role in controlling the nanoprocessing and in optimizing its capabilities. Dr. Hüttmann has shown that the interaction of pulsed irradiation with molecules may lead to complex changes of

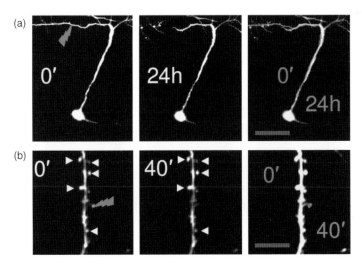

Figure 2   Single spine (b) or single dendrite (a) ablation in living animal (see Figure 2.5 and related text).

the photophysics (fluorescence, thermalization of the excitation) and photochemistry (Chapter 3). In contrast to low irradiance CW-irradiation, when the effect is determined only by the ground state absorption and the radiant exposure, the effect of pulsed irradiation additionally depends on the irradiance, irradiation time, pulse separation, and absorption cross-sections and lifetimes of the excited states. The effects of pulsed irradiation in biomedical applications are diverse, including reduced efficacy of fluorescence formation and photochemistry, increased photobleaching, and the excitation of higher energy levels by simultaneous absorption of more than one photon. As a nonlinear effect, multiphoton absorption confines the interaction of light with molecules to the focal volume and has found considerable applications in microscopy and cell surgery.

Despite the complexity of excited state photophysics, the basic processes relevant for biomedical applications are bleaching of the ground state absorption, stimulated emission from the S_1, excited state absorption from the S_1 and T_1 state, and multiphoton absorption. Numerous experimental studies suggest that excited state absorption reduces fluorescence yield and increases photobleaching. The photophysics and photochemistry are changed dramatically by tailoring the time course of irradiation to the photophysical properties of the chromophore. Successful strategies for reduced photodamage and increased total fluorescence output per molecule have been developed.

This kind of study will bring new optimizations with regards to techniques using these nonlinear interactions. STED (stimulated emission depletion), for example, is a new microscopic imaging strategy for increasing the resolution beyond the Abbe limit, and is based on a sophisticated use of pulsed laser interaction.

In addition, the reduction of photobleaching is of enormous interest in linear and nonlinear laser microscopy.

Multiphoton fluorescence imaging, for example, that uses direct excitation with two or more photons, is now widely used in microcopy. Although in thick samples it profits from the restriction of the excited volume to the thin layer that is actually imaged, photodamage in the focal volume still severely limits imaging speed and/or the length of image sequences in time-lapse microscopy. Multiphoton imaging uses $GW\,cm^{-2}$ to $TW\,cm^{-2}$ irradiance at a wavelength at which the tissue has no ground state absorption.

Adapting the irradiation pattern to chromophores is used to avoid excited state absorption and should reduce photodamage. The continuous development of laser sources and fast scanning devices will give access to a wider range of wavelengths, pulse width, and repetition rate. Together with a growing understanding of the excited state photophysics, a reduction of photodamage in microscopy is expected.

A clear example where photophysics plays a tremendous role is the chromophore-assisted laser inactivation (CALI) technique developed by Professor Daniel J. Jay (Chapter 4).

When CALI was developed there were very few loss-of-function strategies available to address cellular function. Since then, interference RNA strategies that decrease expression have been developed that are facile and high throughput. CALI provides the localized and acute loss of function not provided by RNAi approaches, and thus complements RNAi. It has the following advantages: it is not subject to genetic compensation, is more similar to drug effects in that loss is acute inhibition and not chronic loss of expression, and it is possible to target isoforms and posttranslational modifications, for example, with phosphospecific antibodies. Subcellular micro-CALI and its application may prove valuable in addressing the complex networks of interactions such as those found in dendritic spines in neural connections. In addition, the temporal resolution of CALI may be useful for addressing developmental timing and for addressing genes whose loss results in lethality before the events to be studied.

The unique advantages of CALI may be combined in the future with other recent technological advances. For example, high-throughput approaches combined with molecularly encoded photosensitizers would permit full exploitation of the fusion of molecular genetics with light-based inactivation. Also, combining micro-CALI with high-resolution molecular imaging and single molecular behavior in cells would enhance approaches to understanding subcellular events. The array of photosensitizers now available makes multicolor-CALI possible using different chromophores and different excitations to assess the role of different proteins in a given pathway and potentially the epistasis of their roles. The use of confocal and multiphoton microscopes for CALI and the use of high efficiency photosensitizers provide the possibility of the widespread use of micro-CALI throughout the biomedical research community.

The use of CALI opens up incredible applications in cell biology. Optical knock-out, performed with CALI, will probably constitute a valid alternative to biochemical knock-out. The use of multiphoton processes of interaction will allow the selection of a well-defined spatial region of inactivation of proteins in the cell, in a well-defined temporal window. Such optical control will be useful to study not only some biological processes but it will also permit the optimization of some biochemical and biotechnological protocols, which will open new possibilities in biochemical treatment of cells or the control of their pathway.

The possibility of using photochemistry to control a chemical reaction and the conformation of proteins has also opened up new perspectives in optically manipulating the fluorescence emission state of some chromophores, allowing a new class of chromophores to be developed: the photoswitches (Chapter 5). Dr. Andrew A. Beharry and Professor G. Andrew Woolley have stated in this book that the development of natural photoswitches as tools seems likely to grow quickly as new light sensitive modules are discovered via genome sequencing efforts. Directly co-opting a natural photoswitch for use in a particular case may be possible through imaginative use of chimeras but inevitably will require minor or perhaps major protein engineering efforts. The power of protein engineering to modify (either through structure based design or directed evolution) the properties of light sensitive proteins is considerable, as evidenced by successful efforts with fluorescent proteins. One advantage of the synthetic photoswitch approach is that the properties of the chromophore can often be predicted based on empirical rules or direct calculations. If the structure of the target protein is known then photocontrol of that structure may become possible in a more direct manner via a synthetic approach. Finally, as Kramer and colleagues have pointed out, there are numerous cases where genetic manipulation is simply not possible, so that adding small molecules targeted to specific sites, as with the photoswitchable affinity label (PAL) approach, may be the most effective alternative.

This new class of chromophores constitutes a tremendous input for development of a new class of experiments and microscopy techniques.

Many super-high-resolution techniques, such as two-photon STED [1], STORM [2], and RESOLFT [3], will take advantage of this new class of chromophores. The field of photoswitching is probably one of the more important frontiers of development in super-high-resolution microscopy.

Another application of photochemistry, used to optically manipulate the activity or the conformation of molecules, regards optical neural stimulation, obtained by means of the optical control of membrane potentials. This is obtained with a photochemical process operating on an ionic channel that controls the membrane potential, as described in Chapter 6 by Dr. Rajguru, Dr. Matic and Professor Richter

More recently, laser–neuron interactions have been studied that were not attributed to photochemical reactions by using a HeNe laser to irradiate snail ganglia. The interactions were only able to modify the rate of action potentials in spontaneously active neurons and were not able to elicit action potentials in silent neurons. The authors noted that the temperature effects of irradiation were responsible for the outcomes of the experiments.

As Professor Richter states, the use of lasers in stimulation of neurons and other excitable cells is of increasing interest, especially given the success of pulsed and continuous lasers in exciting neurons. A potential application of pulsed laser light is in neuroprostheses and it may be advantageous in several ways over traditional electrical stimulation. Optical stimulation yields artifact-free signals, which enables the simultaneous recording of neural responses. This is difficult with electrical stimulation and will allow for long-term evaluation of the state of the tissue. It is also advantageous that light can be delivered in a non-contact manner.

One of the main questions remaining unanswered regarding optical stimulation of neural tissue is the mechanism of stimulation. Laser–tissue interactions can occur by various biophysical mechanisms, including photochemical, photothermal, photomechanical, and electric field effects.

In any case, without doubt, the use of photo-activators will have tremendous applications both in biology and medicine. Neuroprostheses using any form of optical stimulation will be implanted in patients and are expected to remain in operation for many years. Therefore, optical neural stimulation must be optimized by establishing the parameters that allow safe, chronic stimulation of the tissue of interest without causing significant damage to neurons. To date, only limited data are available for providing safe optical stimulation over long durations. Further safety studies are required to select safe optical wavelength, energy, size, and placement of the optical source (fiber) for stimulating neurons.

Studies performed at the cellular level in order to understand the mechanism of optical manipulations, as discussed previously, suggest that similar mechanisms are involved in more complex systems like tissues. Here, it is important to underline that the imaging capabilities play a tremendous role in imaging single cells also in depth. For this purpose, techniques devoted to increasing the contrast in tissue imaging are becoming an essential tool to study many aspects of cell behavior, both in pathologies and/or under optical manipulation conditions, such as laser therapies.

In this context, Professor Tuchin's group has demonstrated some specific features of tissue optics and light–tissue interaction (Chapter 7). They show that the administration of optical clearing agents allows one to effectively control the optical properties of tissues. This leads to the essential reduction of scattering and therefore causes much higher transmittance (optical clearing) and the appearance of a large amount of least-scattered and ballistic photons, allowing for the successful application of different optical imaging and spectroscopic (optical biopsy) techniques for medical purposes.

The immersion technique has great potential for noninvasive medical diagnostics using reflectance spectroscopy, OCT (optical coherence tomography), confocal, non-linear, polarized microscopy, and others methods, where scattering is a serious limitation. Optical clearing can increase the effectiveness of several therapeutic and surgical methods using laser beam action on a target area hindered by a depth of tissue.

For this purpose, photodynamic therapy (PDT) is one of the most popular examples of optical manipulation of tissues, where the laser source is used to create photochemical effects affecting targeted cell viability.

As stated by Professor Engels (Chapter 8), PDT as a treatment modality is becoming increasingly understood and accepted, and as treatment parameters, photosensitizer, and dosimetry are developing so new indications may emerge.

The success of PDT for the treatment of solid tumors might very well critically depend on the use of adequate dosimetry as provided by real-time treatment monitoring. Here, the use of PDT as one part of a combination therapy should be pointed out. Many modalities for the treatment of malignant diseases rely on a multimodality approach, with the intention of improving treatment outcome and limit side effects. For PDT, combinations have been identified, for example, the parallel administration of antiangiogenetic and/or anti-inflammatory drugs, that provide synergetic effects.

Although many problems and challenges remain, as Professor Engels says, promising results in preclinical and clinical settings have been seen in PDT and immunemanipulation. The consequences of immunoadjuvant therapy on the PDT light and photosensitizer dosimetry is, however, yet unknown. Future studies will assess the possibility of employing lower total light and drug dosages, as well as the potential for improved selectivity when also activating the adaptive immune system.

A very interesting approach is the treatment of malignant tumors with photochemical internalization (PCI). This concept, developed by Berg *et al.*, employs sensitized vesicles that can enter cells via endocytosis [4, 5]. These vesicles are exposed to activating light, whereby the photodynamic action leads to damage of the photosensitized membrane and their content can be released (or internalized) into the cytosol of the cell. The vesicles could be loaded with a toxic agent, for instance bleomycin, thus selectively destroying irradiated cells. This allows a highly selective chemotherapeutic treatment, potentially resulting in much less side effects.

Notably, two-photon processes can also be exploited in PDT. Here, localization of the interaction process in the focal volume due to the nonlinear interaction allows a well-defined irradiated area.

For multiphoton PDT, light and photosensitizer are delivered over extended time periods. Owing to the extended time consumption, this treatment modality might be more attractive for patients already requiring long and costly hospitalization periods. Technically, small portable light and drug sources need to be implemented and the requirements of accurate light dosimetry need to be elucidated.

There is still strong interest in, and need for, new and better PDT photosensitizers. In the drug development process it has also become evident that it is difficult to identify one chemical molecule possessing all the optimal properties for a PDT treatment. Rather, one has to exploit more complex structures that can provide the possibility of adjusting certain functions independent of the photosensitizer agent.

Another interesting and recent development towards increased drug selectivity is the use of activatable photosensitizers, also referred to as photosensitizer beacons. These have a configuration resulting in inactivation of the photosensitizing ability by a molecular quenching mechanism. This molecular configuration alters once the complex binds to a specific target molecule, where it changes configuration and the quenching is destroyed. This means that the photosensitizer is only efficient once it is bound to a specific target in the tissue. These approaches offer improved selectivity as compared to most currently employed photosensitizers and hence would pose less severe demands on accurate light dosimetry.

There is a pronounced development towards real-time/online PDT-treatment monitoring. Spectroscopic techniques are often employed for these measurements, where the use of small diameter fiber based probes limits invasiveness. This technique may become important for the development of PDT for bigger tissue volumes requiring many fibers for an efficient light delivery.

The notion of real-time treatment monitoring is attractive also when applying an implicit dosimetry model or a dose model based on the biological response to the PDT treatment.

Notably, some imaging techniques will be applied to assist the laser treatment or to study its effectiveness. OCT might be employed to monitor scattering changes during PDT or combined with fluorescence measurements to allow more quantitative photosensitizer monitoring, Doppler OCT or optoacoustic tomography might be employed before, during, and after PDT to assess vascular shutdown. Finally, it should again be emphasized that the possibility offered by realtime dosimetry and/or combination strategies to diminish treatment related side effects as treatment parameters can be individualized, and/or each treatment modality must not be used to any extreme doses.

Another technique of laser manipulation of tissues is that described by Dr. Pini for welding applications (Chapter 9). In this procedure, great importance is represented, however, by an exogenous chromophore enabling the welding operation. As stated by Dr. Pini, current efforts to extend the range of clinical applications of laser welding include, in particular, the development of novel concepts of exogenous chromophores.

Conventional exogenous chromophores in common use in surgical practices are organic dyes, such as the ICG. Despite the experimental and clinical achievements, these molecules are less than ideal and exhibit poor performances in several critical respects. For instance, the light extinction efficiency and photochemical stability of solutions of organic dyes are limited.

Along with other drawbacks, these limitations hinder, for example, better control over perfusion and delivery to the biological targets of interest, such as, for example, topically through biocompatible insets and systemically through intravenous injection. Therefore, there is growing interest in the introduction of alternative concepts of exogenous chromophores that may improve the feasibility of laser welding and other kinds of diagnostic and therapeutic applications.

Recent progress in the design and synthesis of novel colloidal suspensions has produced various nanoparticles with excellent optical response, which hold the potential to outclass the organic dyes presently in use. Examples include, hybrid materials composed of ICG molecules embedded within organic or inorganic nanoparticle. These systems rest on the ordinary concepts of photothermal conversion, realize high local concentrations of the organic dyes, exhibit better stability in the biological environment, and may become conjugated with further functional molecules.

Beside these hybrid materials, there are alternative proposals of exogenous chromophores that rely on radically innovative concepts of photothermal conversion. Here we mention metal oxide and metal nanoparticles with magnetic and plasmonic resonances, respectively.

On the other hand, as one further perspective, the wealth of concepts available to modify the surface of metal nanoparticles may enable, for example, their inclusion into durable and biocompatible films and gels, which may be loaded with additional drugs and factors to optimize delivery of the exogenous chromophores and the repair after the surgery.

Currently, the replacement of organic dyes with laser activated nanoparticles is an inspiring possibility – one that is still quite far from the clinical arena. As a very recent proposal, there is limited experimental proof of its actual potential in the laser welding of connective tissues. Additional aspects of practical concern include the biocompatibility and sustainability of laser activated nanoparticle. In short, the new and thriving context of nanomedicine is quickly penetrating laser welding technology as well, and is certainly among its most pioneering frontiers.

Industrial developments in this sense have followed and are now allowing more and better performing solutions. New optical tweezers manipulators, laser or even LED based sources have been used in new light-based therapies for tissue application in the field of urology, neurosurgery, dermatology, ophthalmology, and so on. Imaging assisted operation is becoming increasingly important for tumor treatment, both for therapy and surgery operations. Moreover, optical control of nerve reaction and stimuli, based on implanted neuro-chips, will assist prothesis implantation and control or, even further, control the symptoms in some neurological pathologies.

In conclusion, all the scientific and technological developments have shown how the capability of imaging together with the possibility to optically manipulate the sample has created a growing area of applications in the field of cell biology. Future developments are hard to predict at present since the area is growing much faster than expected. Certainly, we can expect in the near future a new frontier in medical and biology fields in the study of many phenomena and pathologies, and possibly new kinds of therapies.

References

1 Moneron, G. and Hell, S.W. (2009) *Opt. Express*, **17**, 14567–14573.

2 Rust, M.J., Bates, M., and Zhuang, X. (2006) *Nat. Methods*, **3**, 793–795.

3 Schwentker, M., Bock, H., Hofmann, M., Jakobs, S., Bewersdorf, J., Eggeling, C., and Hell, S.W. (2007) *Microsc. Res. Tech.*, **70** (3), 269–280.

4 Berg, K., Dietze, A., Kaalhus, O., and Høgset, A. (2005) Site-specific drug delivery by photochemical internalization enhances the antitumor effect of bleomycin. *Clin. Cancer Res.*, **11** (23), 8476–8485.

5 Berg, K., Folini, M., Prasmickaite, L., Selbo, P.K., Bonsted, A., Engesaeter, B., Zaffaroni, N., Weyergang, A., Dietze, A., Maelandsmo, G.M., Wagner, E., Norum, O.-J, and Høgset, A. (2007) Photochemical internalization: a new tool for drug delivery. *Curr. Pharm. Biotechnol.*, **8** (6), 362–372.

Index

a

absorption   115ff, 122, 127, 128, 133, 139, 146ff, 153ff
– bleaching   56
– centers   128
– coefficient   58, 115, 117, 120, 122, 124, 127, 136
– excited state   56
– multiphoton   55, 60
– spectrum   116, 117, 128, 147, 149
– triplet–triplet   58
– two-photon   60
adipocytes   24
albumin   205ff
anastomosis   210ff
Atto532 dye   65
autofluorescence   14
azobenzene   85ff

b

bio-heat equation   213
biomaterials   204
biopolymers   206
bleaching   63, 65
– transient   62, 64
blood vessel   44
Bunsen–Roscoe law   51

c

caged compounds   83
caged molecule   102
CALI, *see* chromophore-assisted light inactivation (CALI)
central nervous system   36, 42ff
channelrhodopsin   89, 103
channelrhodopsin-2   91ff
chitosan   206
ChR2   89

chromophore-assisted light inactivation (CALI)   71ff
– alkaline phosphatase   72
– calcineurin   76
– environmentally sensitive fluorophores   73
– fasciclin I   72
– β-galactosidase   72
– gold nanoparticles   73
– half-maximal radius of damage   73, 74
– induced free radical generation   74
– IP3 receptor   76
– photothermal mechanism   74
– pp60c-src   76
– RNA aptamers   73
– small molecule binders   73
chromophores   100, 203
– endogenous   203
– exogenous   204
chromosomes   38ff
cochlear implants   99
collagen   14, 205
– denaturation   207ff
– relaxation   208
– shrinkage   208
connective tissues   207
cornea   205ff
CWLW, *see* laser welding, continuous wave

d

dark states   65
deep brain stimulators   99
differentiation   19
– adipogenic   28
– chondrogenic   28
– osteogenic   28
diffusion   123, 125, 130ff, 135, 139, 141, 142, 144, 147, 155
– coefficient   130, 131, 133ff, 138

Laser Imaging and Manipulation in Cell Biology. Edited by Francesco S. Pavone
Copyright © 2010 WILEY-VCH Verlag GmbH & Co. KGaA, Weinheim
ISBN: 978-3-527-40929-7

– model   135
– one-dimensional problem   135
– rate   131

e
ECM   24
Einstein coefficient   56
elastin   14
emission, stimulated   65ff
extracellular matrix   205ff

f
FALI, *see* fluorophore-assisted light inactivation (FALI)
femtosecond lasers   9ff, 35ff
FlAsH-FALI   73
– synaptotagmin I   77
flavins   14
fluorescence-assisted cell sorting (FACS)   19
fluorescence lifetime imaging (FLIM)   17
fluorophore-assisted light inactivation (FALI)   72, 74
– amino acyl side chains, modification   74
– β-galactosidase   74
– hsp90 alpha   77
– neuropilin   1 77
– pKi-67   74
– poliovirus receptor (CD155)   77
free electron   36, 37ff

g
GFP, *see* green fluorescent protein (GFP)
GFP-CALI, capping protein (CP)   76
glucose   123ff, 129, 134, 139ff, 143, 145ff, 151ff
– monitoring   145ff
– sensing   115, 146
glycosaminoglycans   206ff
green fluorescent protein (GFP)   27, 38, 42ff, 65, 72

h
HA, *see* hyaluronic acid
halorhodopsin   89
heat damage, *see* thermal damage
hematoporphyrin derivative (HPD)   62
histology   45ff
hyaluronic acid   220
hydrogen   53

i
ICG, *see* indocyanine green (ICG)
immersion   115, 122, 124, 125, 130, 136, 138, 139, 145, 150ff, 154, 155

– agents   136, 138
indocyanine green (ICG)   205
internal conversion   55
intersystem crossing   55
ion channels   86ff, 107
ionization, multiphoton   11, 60
ionotropic glutamate receptor (iGluR)   86, 88ff
irradiance   52
irradiation, pulsed   51

j
Jablonski diagram   53ff

k
keratoplasty   215ff
KillerRed   73
– crystal structure   75
knock-out   30

l
laser   99
– ablation   71
– axotomy   42
– diode   205ff
– near-infrared   103, 205
– pulsed near-infrared   100
– soldering   205
laser-neuron interactions   103
laser-stimulated neurons   103
laser-tissue interactions   106
laser welding   203ff
– continuous wave   214
– hard   209
– mechanism   206
– moderate   209
– ophthalmology   215
– pulsed   214
– soft   209
– surgical applications   214, 215
– temperature control   211
– vascular surgery   219ff
light pulses   103
LOV   92, 93ff

m
magnetic moment   53
malachite green   71
melanin   14
micro-CALI   75
– calcineurin   75
– dynactin   75
– ezrin   75
– IP3 receptor   75

– talin   75
– vinculin   75
microscopy
– confocal   61, 145, 149, 153
– confocal laser   63
– fluorescence   10, 61, 63
– laser scanning   63
– multiphoton   10
– polarized   152, 155
– second-harmonic generation (SHG)   149, 150, 152
– time-lapse   67
– two-photon   10, 151, 152
microtubule   38, 39ff
mitochondrion   40, 41ff
mitotic spindle   38ff
modes
– magnetic   53
– rotational   53
molecules, electronic states   53
multi-photon fluorescence microscopy   35
multiphoton microscopy   44ff

n
NAD(P)H   12ff
nanoparticles   224ff
nanoprocessing   13
nanosurgery   18
natural photoswitches   89, 93ff
near-infrared radiation   104
– laser radiation   108
neural prostheses   99
neural stimulation   104, 105, 107
– with optical radiation   100
neuronal   41ff, 46ff
neurons   35ff, 41ff, 102, 106
neuroprostheses   106, 108
NpHR   89ff
numerical aperture   61

o
objective   45ff
objective lens   36ff
optical breakdown   36ff, 39ff, 61
optical cleaning   27
optical clearing   122, 123, 125, 126, 128, 129, 132, 144, 146, 149ff
– agents (OCAs)   115, 122ff, 130ff, 142, 143, 149, 151, 154
optical coherent tomography (OCT)   133, 138, 139, 141ff, 147, 155
– amplitude (OCTA)   139ff, 144
– signal slope (OCTSS)   139ff
optical histology   43

optical projection tomography (OPT)   153
optical radiation   108
optical stimulation   104, 106, 108
– of neural tissue   103
optoporation, laser   28

p
Pauli principle   53
photoacoustic calorimetry   74
photochemical   107
– mechanism   101
– reactions   103
photodynamic therapy (PDT)   61ff
photomechanical   107
photomultiplier   37
photon
– energy   53
– momentum   53
photophysics   52ff
photostimulation   99, 103
photothermal infrared nerve stimulation   107
phytochromes   91ff
plasma formation   61
PLW, *see* laser welding, pulsed
protein crosslinking   74
proteoglycans   206ff
pulsed infrared lasers   104, 105

q
quantum yield   59

r
radical anion   66
radiometers   212
Raman scattering, coherent anti-stokes   55
Raman spectroscopy   133, 134, 148, 149
rate constants   56ff
rate equations   56ff
Rayleigh criterion   65
reaction, photochemical   62
ReAsH CALI   73
refractive index   115, 116, 120, 122, 123, 126ff, 137ff, 147, 153
– complex   118, 127
– matching   123, 126, 127, 133, 150
– mismatch   120, 126, 144, 146, 147
– relative   129
repetition rate   36ff, 39ff, 46ff
resolution, diffraction limited   65
rhodamine 6G   64
rhodopsin   89, 91ff
– ion channel   103

s

saturation   58, 64
scanning head   37
scattering   115ff, 122, 124ff, 133, 135, 137ff, 142, 144ff
– anisotropy factor   120
– centers   126ff
– coefficient   115, 118, 120, 122, 124, 126, 136, 137, 139, 147
– – reduced   118, 120, 128, 147
– – transport   129
– Mie theory   154
– multiple   117
– particles   116, 120, 137
– single   126
– spectrum   128
– tissues   35
second-harmonic generation (SHG)   11
sensitizer   62
singlet states   54
solders   204ff
SPARK   88
spectral imaging   20
spheroids   26
stem cells   9ff
– pancreatic   25
– salivary gland   25
– therapy   30

stimulated emission depletion (STED)   66
stimulation   108
stretching   122
super-resolution   65
synthetic photoswitches   84

t

thermal activation   55
thermal damage   204
thermalization   59
thermocamera   212
time-correlated single photon counting (TCSPC)   17
tissue coagulation   122
– compression   122
– dehydration   115, 122ff, 133, 146, 150, 151, 155
tomograph, multiphoton   12
transitions
– non-radiative   55
– radiative   55
triplet states   54

w

wavefunctions   53
welding, photochemical   204
wounds   203ff